Jacquard's Web

Jacquard's Web

How a hand-loom led to the birth of the information age

by James Essinger

OXFORD
UNIVERSITY PRESS

OXFORD
UNIVERSITY PRESS

Great Clarendon Street, Oxford OX2 6DP

Oxford University Press is a department of the University of Oxford.
It furthers the University's objective of excellence in research, scholarship,
and education by publishing worldwide in

Oxford New York

Auckland Bangkok Buenos Aires Cape Town Chennai
Dar es Salaam Delhi Hong Kong Istanbul Karachi Kolkata
Kuala Lumpur Madrid Melbourne Mexico City Mumbai Nairobi
São Paulo Shanghai Taipei Tokyo Toronto

Oxford is a registered trade mark of Oxford University Press
in the UK and in certain other countries

British Library Cataloguing in Publication Data
Data available

ISBN 0192805770

Typeset in Perpetua by
Footnote Graphics Limited, Warminster, Wilts
Printed in Great Britain by
Clays Ltd., St Ives plc

To my parents Mary and Ted,
and in memory of Julius, Rega, and Uli Essinger

Many ingenious minds labour in the throes of invention, until at length the master mind, the strong practical man, steps forward..., applies the principle successfully, and the thing is done.

Samuel Smiles, *Self-Help,* 1862

Contents

Contents

Illustrations

Illustrations

A LA MÉMOIRE DE J. M. JACQUARD.

Né a Lyon le 7 Juillet 1752. Mort le 7 Aout 1834.

·1·
The engraving that wasn't

The portrait of Jacquard was, in fact, a sheet of woven silk, framed and glazed, but looking so perfectly like an engraving, that it has been mistaken for such by two members of the Royal Academy.

Charles Babbage,
Passages from the Life of a Philosopher, 1864

If you wanted to be part of the scientific and literary set in the London of the 1840s, you would have done just about anything to beg, steal, or borrow an invitation to one of Charles Babbage's famous soirées. Charles Babbage was a scientist, philosopher, engineer, mathematician, and writer. He had devoted much of his life to trying to build two types of calculating machines made out of cogwheels. He called these the Difference Engine and the Analytical Engine. So far he had failed to complete either of them, yet his struggle to do so had won him admiration even from people who were convinced his efforts were doomed to end in failure.

Babbage's friends saw him as fascinating and brilliant. His enemies tended to think him moody, temperamental, and over-

---

(*left*) The Jacquard portrait.

rated. But even those who found him more irascible than inspired had to admit he gave splendid parties. These were held at his home: Number One Dorset Street. This is close to Manchester Square in the London district of Marylebone. Babbage's soirées took place on Saturdays during the 'season', the time of the year when fashionable society would attend a whole range of gatherings, dinners, and balls. The season usually lasted from late March until the end of July.

Babbage had moved to Dorset Street in 1828. For the first few years his parties there were private functions for family and close friends. But in the early 1830s he broadened the list of guests to include many of the leading luminaries of British intellectual life. During the next decade his social events became renowned throughout the capital. They frequently lasted until well after midnight, under the glow of thousands of candles. Three hundred guests, or even more, might attend. Invitations were so prized that many of the most famous people in London used to write begging letters to Babbage to try to secure an invitation for themselves, their family, or friends.

The soirées eventually became one of the great rendezvous points for liberal intellectuals in Victorian London. Charles Dickens, Charles Darwin, Lord Byron's daughter Ada Lovelace, the actor William Macready, the scientist Henry Fitton and his wife, the geologist Charles Lyell, the self-taught mathematician Mary Somerville and her family, the anatomist Richard Owen, the magistrate William Broderip, the astronomer Sir John Herschel; these are just a few of the 'names' who were often to be found at Babbage's parties. An account by the American man of letters George Ticknor of a visit to one of Babbage's parties on 26 May 1838 gives us a glimpse of what they were like:

> About eleven o'clock we got away from Lord Fitzwilliam's and went to Mr Babbage's. It was very crowded tonight, and very brilliant; for among the people there were Hallam, Milman and his pretty wife; the Bishop of Norwich, Stanley, the

Bishop of Hereford, Musgrave, both the Hellenists; Rogers, Sir J. Herschel and his beautiful wife, Sedgwick, Mrs Somerville and her daughters, Senior, the Taylors, Sir F. Chantrey, Jane Porter, Lady Morgan, and I know not how many others. We seemed really to know as many people as we should in a party at home, which is a rare thing in a strange capital, and rarest of all in this vast overgrown London. Notwithstanding, therefore, our fatiguing day, we enjoyed it very much.

Babbage delighted in entertaining the guests who came to his soirées with ingenious devices and gimmicks. In 1832, having after a decade's struggle finally managed to build one-seventh of the mechanism of the Difference Engine, Babbage proudly mounted the completed portion of the machine in a case of mahogany and glass. For eight years it was the most prominent conversation piece at his glittering events.

Then, in the spring of 1840, Babbage started exhibiting something else.

On the face of it, this new exhibit was nothing more than an unassuming portrait of an inventor in his workshop. The portrait shows the inventor sitting in a luxurious cushioned chair at his work bench. He is holding a pair of callipers against long strips of cardboard that have tiny holes punched in them. The bench also accommodates a model of a loom. Hanging upon a rack on a wall behind the inventor are chisels and other tools in a variety of shapes and sizes. Rolled-up plans are poking out of a drawer on a table beneath the rack.

The portrait gives the impression of being an informal snapshot of the inventor as he momentarily turns away from his work and glances at the artist. He has a thoughtful, frowning air about him, and his well-cut coat and general air of prosperity suggest that this is an inventor who has enjoyed some success.

Anybody giving the portrait a cursory inspection would assume that it is an engraving. This is exactly what the vast majority of Babbage's guests thought when they first set eyes on it. Babbage, a wily old fellow who took as much delight in exposing the errors and folly of his friends as he did in advancing the cause of science, enjoyed showing the portrait to his guests. He would then ask them how they thought it had been made. When they told him they were sure it was an engraving, as they usually did, he gave a knowing smile.

One evening in 1842, two of the most distinguished people in the realm attended a soirée. They were the Duke of Wellington and Prince Albert, Queen Victoria's husband. The 'Iron Duke' was the hero of Waterloo and a former Prime Minister. Prince Albert was famous for his intellect and for the important, even essential, role he played in governing Britain. Officially he had no power, but in practice the Queen deferred to his judgement and opinion on almost every matter. She usually succeeded in persuading her ministers to do the same.

Almost as soon as the Duke and the Prince arrived, Babbage showed them the portrait. The Prince asked Babbage why he thought the portrait so important. Babbage replied, in characteristically enigmatic fashion, 'It will greatly assist in explaining the nature of my calculating machine, the Analytical Engine.'

Once the two guests had examined the portrait, Babbage asked them what they thought it was. The Duke of Wellington, getting things wrong for once, made the usual mistake of responding that it must be an engraving. But it turned out the Prince knew the truth, having apparently heard of the portrait before. He informed the Duke of Wellington that the portrait was not an engraving at all, but a *woven piece of fabric*.

And the Prince was absolutely right. The picture was, in fact, a woven silk image. It depicts a French inventor named Joseph-Marie Jacquard. He had died six years earlier, on 7 August 1834. It was Jacquard who had invented the very loom on which the portrait was woven. This, the Jacquard loom, was the world's first automatic machine for weaving elaborate and beautiful images into silk.

The portrait, deliberately designed to illustrate and show off the Jacquard loom's capabilities, is so complex it contains 24 000 rows of weaving. Every single row was controlled by what was in effect an early nineteenth-century programming device—a punched card. The 24 000 cards gave the loom precise instructions for weaving the portrait. These punched cards lie at the heart of Jacquard's brilliant concept of an automatic loom that weaves complex patterns and images.

The portrait was woven in Lyons in 1838 by a weaving firm named Didier Petit & Co. It was based on an oil portrait of Jacquard originally painted by a Lyons artist, Claude Bonnefond, at the time the director of the City's school of Fine Arts. Bonnefond took care to give the punched cards a prominent place in the portrait.

A few originals of the woven portrait still exist today. There is one in the reserve collection of the Science Museum in London, although unfortunately the portrait is no longer on general display, for the Museum has temporarily slimmed down its history of computing exhibition as a preparatory step to planning a major new display on computing and communications. However, you can ask for permission to view the Jacquard portrait in the reserve collection.

As you gaze into Jacquard's stern features, it is difficult to believe that this faded, rather small picture (it only measures 20 by 14 inches), can be an illustration of a technology, developed more than two centuries ago, that was to alter our world beyond recognition.

Yet who exactly *was* Joseph-Marie Jacquard? How did he come to invent a loom that could weave pictures? And how did his extraordinary idea lead to the global information revolution that is continuing to transform the world in which we live today?

·2·
A better mousetrap

To call forth figures, flowers, or patterns of any other kind,
different means are necessary.

Dionysius Lardner,
A Treatise on the Silk Manufacture, 1831

Our story begins with the discovery of silk. And, according to
legend, the story of silk begins with a cup of tea.

One afternoon, sometime around the year 2700 BC, the Chinese empress Si Ling-Chi was strolling around her garden. She
chanced to pick a fuzzy white cocoon from her favourite mulberry tree. Taking it over to her tea-table, she started toying
with it. Her fingers were clumsy; a moment later she accidentally
dropped it into her hot cup of tea. Fishing it out, she discovered
to her surprise that she could pull out a long strand of thread
from the cocoon. She pulled and pulled; the strand grew longer
and longer.

Si Ling-Chi had discovered silk. A natural fibre produced by
most spiders and by many caterpillars, silk is created in particular
abundance and strength by the silkworm, which is really a cater-
pillar and whose diet consists almost entirely of mulberry leaves.

Whether this story of the origin of silk is literally true or a myth concocted to flatter the empress does not really matter. What is certain is that the properties of silk, and the secrets of its production, were first discovered in China about three thousand years ago.

Astonishingly, the average silkworm cocoon contains a single thread of silk that when fully unwound is often as much as a full kilometre in length. The discovery of silk, and the revelation that numerous threads could be unrolled from a cocoon, spun into a stronger thread and used to make a fabric, caused a sensation. Never before had anyone known of a fabric so soft, strong, durable, as readily dyed, and yet so dirt-resistant. Silk appeared the ultimate luxury. Indeed, it almost seemed a minor miracle.

The Chinese zealously guarded the source of silk and its production for many centuries. They didn't even start exporting silk until more than two millennia after they first discovered it. But, in around 140 BC, the secret of silk-making spread to India as an inevitable result of the thriving trade in silk between that country and China. Yet more than six centuries passed before the mystery of silk production finally reached Europe, which had believed for many centuries that silk thread was simply picked from certain trees native to China.

In AD 550, two Persian monks who had lived as missionaries in China related the truth about the origins of silk to the Byzantine emperor Justinian I. A man who was used to being obeyed, Justinian ordered the monks to return to China and smuggle silkworm eggs to Constantinople. He also promised them a hefty financial reward if they were successful. The monks, now pretending to be missionaries when they were in fact engaged on the emperor's mission, trekked back to China and did his bidding, concealing thousands of silkworm eggs in the hollows of their bamboo canes.

Silkworm eggs enter a state of suspended animation when kept in the cold and dark. Once the eggs reached Constantinople, they were placed on top of a pile of warm manure and in the sunshine. In these favourable conditions the eggs soon hatched. The emerging larvae were fed on leaves of the wild mulberry that grew abundantly in and around the city. The hardy silkworms resulting from this cunning act of industrial espionage gave birth to the European silk-making industry.

Homo sapiens is believed to have first evolved about 200 000 years ago. For almost all of this time, the new species clothed itself in skins and furs torn from the bodies of animals that had been slaughtered for food. The production of fabric began, by comparison, far more recently, at about the same time as recorded history. This is no coincidence: the writing of history and fabric-making both require a considerable sophistication of civilization.

The first plant used for cloth was flax, still employed extensively today to make linen. The earliest surviving fragments of linen are found in Egypt. They were made there in about 4500 BC. The production of linen depended on a major new invention: weaving.

Weaving is the interlacing of two or more sets of strands of fibre at right angles to each other to form a useful material. It's a simple process when rigid fibres such as reeds are being used. In this case, as anyone who has done raffia work will know, the weaving can easily be undertaken manually. But rigid fibres don't make comfortable clothing: this needs to be produced from soft, flexible yarns and can only be woven in a neat and convenient way by making use of a loom.

A loom is designed to hold a set of horizontal threads— known as the warp—so that they do not tangle. The loom must also facilitate the weaving process by allowing a set of vertical threads—the weft—to be interlaced with this warp. Most looms only have a few basic parts. Usually one end of the warp is tied to a *warp beam* (also known as the back beam or warp bar) while the other end is fastened to the *cloth beam* (also known as the front beam) on which the finished fabric can be rolled.

This kind of hand-loom is well-suited to weaving plain, undecorated fabric, but it cannot weave more ambitious kinds of fabrics that contain complex patterns or—most ambitious of all—*images*. Fabrics containing complex patterns or images are known as *decorated* fabrics. They can only be woven by a loom that allows the raising and lowering of individual warp threads to permit the different coloured weft threads to be inserted by the shuttle in such a way that a design can be created in the fabric.

The first loom that made it possible to create a pattern in fabric was called a *drawloom* because the loom allowed the warp threads to be *drawn* up individually to create the design to be woven. The raising of the warp threads forms what is known as the 'shed': the opening made between the threads of the warp to allow the loom's shuttle to pass through.

The first drawloom was invented in China in or around the second century BC. It is no surprise that the original invention of the drawloom should have taken place in China, for the extreme fineness and flexibility of silk fabric makes it ideal for having images woven into it. In such designs the individual silk threads, compacted together by the weaver using a comb-like device, are so fine that they cannot be individually distinguished. The woven fabric containing the design comes out of the loom looking like an emerging oil painting.

But the drawloom, while ingenious, was in fact a highly unsatisfactory apparatus. The big problem was that the arrangement of the individual warp threads was usually different for every single row of weaving. In practice, the more detailed the pattern or image, the more different the arrangement of threads was likely to be for each row. This meant that the drawloom operators had to make decisions on a row-by-row basis about whether an individual warp thread should be raised, or kept in its lowered position. There could easily be up to 500 warp threads in a single row of weaving, and while some of these could usually be lifted *en masse*, the job of working a drawloom remained incredibly laborious. It always required two people—the weaver

A drawloom.

who operated the shuttle, and the 'draw-boy' (typically a boy or a young man)—who stood on top of the loom and raised or lowered the warp. As for the rate of production possible on the drawloom, even the most experienced two-man drawloom teams could only manage a couple of rows of woven fabric each minute. (These rows are known by weavers as 'picks'.) When we consider the laboriousness of the job, it is surprising the duo could even work as fast as this.

The drawloom is really only a bit-part character in *Jacquard's Web*. But it was precisely *because* the drawloom was so maddeningly slow and tedious to use that Joseph-Marie Jacquard was inspired to invent something better.

What Jacquard set out to do was fundamentally simple. His goal was to revolutionize the speed with which the silk-weavers of his home town, the great French city of Lyons, could create the most beautiful decorated silk fabrics the world had ever seen. To achieve this objective, he had to invent a completely new kind of machine: a loom that was capable of being programmed.

The loom would need to be made so that it could weave one particular image, and then, having been given different instructions, be capable of weaving a *completely different* image.

Of course, another way of achieving the same objective would have been to build a different loom for every different design the weaver wanted to weave. For example, if pictures of roses were popular this month Jacquard might have tried to build a loom that only wove roses. The loom would have had to contain inside its mechanism all the instructions for weaving a rose. Such a loom would, in effect, have been a 'dedicated' rose-weaving loom. The concept of a machine devoted to a particular task would have been familiar to Jacquard from his knowledge of the machines fostered by the great Industrial Revolution that was happening in Britain.

But Jacquard, whose father had been a master silk-weaver in Lyons, was only too aware that the problem with building a loom dedicated to weaving, say, an image of a rose is simply this: *what do you do when roses go out of fashion?*

Well, of course you could build another loom, one dedicated, for example, to weaving pictures of a lily or of a Lyons skyscape. But managing matters in this way is expensive, impractical, and a bit silly. After all, by the time you have built your lily-weaving loom, lilies, too, might have gone out of fashion. And even when lilies come back into fashion, the designs popular last time round may not be popular now.

In practice, making a machine that is dedicated to carrying out only one very specific set of complex procedures makes little sense unless you can be *absolutely certain* there will be an indefinite and extensive demand for what that machine can produce. And even if there is, offering the user *flexibility* in what the machine can do is probably a much better idea.

Joseph-Marie Jacquard never had any doubt that the only way of solving the problem of inadequacies of the drawloom was to build a new kind of loom that could be instructed to weave any design at all. The challenge he faced was to find a way to convey the instructions of what to weave to the loom.

'Make a better mousetrap and the world will beat a path to your door.' This saying, attributed to the American philosopher Ralph Waldo Emerson, neatly summarizes the eventual success of Jacquard's quest.

Jacquard's attempt to improve on the drawloom was, in effect, his bid to invent a better mousetrap. His mechanical genius and stubborn persistence were the catalysts he needed. But just as William Shakespeare might never have become a great poet and playwright without the wonderful stimulation and energy of Renaissance London all around him, Jacquard would most likely never have blossomed as an inventor had he not lived in Lyons, in his day the silk-weaving capital of the world.

With a population today of 415 000, Lyons is France's second most heavily populated city after Paris. Ever since Roman times the city has been proud of being second only to the French capital in prosperity, the richness of its cultural life and the ingenuity of its citizens.

Lyons is situated on a confluence of the Rhône and Saône rivers. These divide Lyons in a roughly north–south direction, rather like the twin prongs of a tuning-fork, with the Saône on the west, and the Rhône on the east. Old Lyons lies on the west bank of the Saône. On the other side of this river is a long tongue of land called La Presqu'île, bordered on the east by the Rhône. There is a high plateau at the northernmost extremity of La Presqu'île known as Croix Rousse.

During the Renaissance of the fifteenth century, Lyons experienced a period of economic prosperity and intellectual brilliance. In 1464, the city held the first of many commercial fairs; a development spurred by the recent arrival in the city of Italian merchant bankers. The families who dominated Lyons at this time started to acquire great fortunes from a near-monopoly of the money-lending business. These banker families stimulated trade in both their professional and personal capacities. They extended credit and spent lavishly on establishing and maintaining their lifestyles. New types of crafts developed; foreigners poured into the city. Three other enormous trade fairs were held in Lyons during the fifteenth century.

By the sixteenth century, France's monarchs had started to eye Italy as a target for annexation. They began to make extended visits to Lyons, a city much closer to the Italian border than Paris. They were invariably accompanied by hundreds of courtiers, chefs, doctors, hairdressers, mistresses, nobles, and other servants and hangers-on who comprised their entourage. During the sixteenth century Lyons received twenty such protracted royal missions. It was now truly a Royal City as well as a wealthy one. Its municipal pride had never been greater.

Kings, queens, and others of noble blood thought linen and cotton perfectly good enough for their lowly-born subjects, but absolutely unacceptable to elevated people such as themselves. For *them*, what other fabric could be suitable but the king of fabrics: silk? And there was only one way of producing the fabulously ornate damasks, satins, taffetas, brocades, and lampas (a patterned silk that imitated Indian painted and dyed textiles) which France's royalty and aristocracy took for granted much as we take cotton for granted today. *Every stitch* of the deliciously soft, heavy, decorated silks that clothed their privileged forms during the day, kept them warm during the night, blocked out the disruptive sunlight in their bedrooms, or glorified the walls of their palaces, had to be woven laboriously on a drawloom by a master-weaver and an assistant perched precariously on top of the archaic mechanism.

The introduction of silk production and silk-weaving into France had been initiated by the French king Louis XI, who held the throne from 1461 to 1483. In 1480 he offered expert silk-workers from Genoa, Florence, and Venice valuable incentives to come and ply their trade in the French city of Tours, in the Loire valley. Louis's initiative to found a silk-making industry at Tours was a success, but the French silk industry remained confined there until the reign of François I, King from 1515 to 1547.

Why was Louis so eager to create an indigenous silk industry? One reason was selfish: he loved silk and wanted to secure his own supply of it. But he had an altruistic reason, too. The silk industry was the world's first luxury business, and experience showed that silk production and silk-weaving always brought prosperity to any region where they were practised.

In 1515, François I ascended the French throne. A man who adored conquest, he wasted no time in launching an invasion of Italy. There, he defeated the Duke of Milan, whose army was no match for his French adversaries. After this victory, France owned the Duchy of Milan for fourteen years. Whatever the rights or wrongs of the invasion, it certainly played a key role in developing the French silk industry. François loved silk even more than Louis had. The new king did his utmost to encourage expert Italian silk-workers to set up shop in Lyons. The king personally supervised the development of a silk industry there, making regular visits, offering awards and incentives for successful production and setting up guilds that initiated high standards of professional attainment. Silk-making and silk-weaving spread to other French cities as well as Lyons, but none of them achieved the prominence of the Lyons silk industry.

No one doubted that the second city of France was the perfect place to nurture the country's home-grown silk industry. It was a city the French kings adored, as much because of its warm climate and great beauty as because of its proximity to the coveted duchies and cities of Italy. French royalty could live in Lyons with all the comforts of court while planning their next campaign. And as well as the royal court, there were many noblemen and wealthy people in Lyons who formed a ready market for quality silk fabrics. The city also had a skilled, comparatively well-educated workforce accustomed to working long hours in well-organized weaving studios.

By 1566, at a time when the total population of Lyons was about 120 000, more than one-tenth of these people were silk-weavers. Over the next few decades, the proportion of the weaving population continued to grow relative to the entire population, so that by the end of the sixteenth century, an actual *majority* of people living in Lyons depended either directly or indirectly on the silk industry for their livelihoods. The decorated silk fabrics woven in Lyons were the envy of the world. Even fabrics woven in China, where silk production and silk-weaving had originated, did not compare in quality, beauty, and artistic detail to those produced in Lyons.

The most valuable and elaborate silk fabric of all was brocade. Many examples of brocades woven in Lyons during the sixteenth century have survived and can be viewed today at the Lyons Textiles Museum. Their colours have faded a little, but the artistry continues to amaze even 400 years later. The fabrics do indeed look more like woven oil paintings than cloth.

From the earliest days of the Lyons silk industry, the high price commanded by brocade meant that any improvement, however slight, in the speed of the drawloom could make a major contribution to the profits of a weaving studio. There was consequently every incentive for resourceful master-weavers, or the mechanics they employed, to try to do what they could in this direction.

A number of small but significant improvements to the drawloom were introduced. One of the most important was a sequence of levers allowing entire bundles of weighted warp thread controllers to be lifted in one go. The various refinements increased the speed with which the drawloom could be used. By the start of the eighteenth century, the drawlooms used in Lyons had become as good as they were ever going to get. The weaver and the manipulator of the cords could produce two picks of woven fabric each minute.

Yet the essential nature of the drawloom had not really improved at all. It still required two people to operate it, sometimes even more if the design was really complex. And even with the help of the levers, the warp threads still had to be lifted manually. At the maximum rate of two woven rows a minute, a mere inch of brocade fabric still required a *full working day* to complete. This is because silk fabric is extremely fine, and to create what was in effect a woven painting, the rows of thread had to be compacted extremely tightly together after every pick. At this slow-motion rate of production, a large order, such as a set of curtains for a large room, could take months to weave, even if the customer contracted several weaving studios to handle different parts of the order at the same time.

The real problem was that the drawloom was not a *machine* at all. Instead, it was really only a device for facilitating the *manual* weaving of patterns or images in fabric. Surely there was some other way, and ideally a much faster way, of weaving decorated silk fabric?

What was required was a method of ordering, with complete precision, the lifting of the warp threads that formed the shed in a better way than having a draw-boy doing the whole thing by hand. With enormous financial returns certain to be won by any master-weaver or master-manufacturer who brought in a revolutionary improvement to the drawloom, it is hardly surprising that there was money, and official encouragement, available for inventors whose ideas offered a chance of creating such a machine. By the middle of the eighteenth century, numerous weavers, master-weavers, and even French Government officials were working on the problem.

Before Jacquard made his great breakthrough, the most important pioneer of automated decorated silk-weaving was a civil servant called Jacques de Vaucanson. Appointed inspector of French silk factories in 1741, it took him only four years to come up with a better idea than the drawloom.

The idea of a mechanism that controlled the raising of the

warp threads *mechanically* instead of manually had been put forward by others working on an improved version of the drawloom, but it was de Vaucanson who first made the idea seem feasible. His plan was for a special control box to be situated above the loom. There, it acted directly on hooks fastened to the cords that controlled the raising warp yarns. The hooks passed through needles and were raised by a strong metal bar. The needles were selected by a mechanism based around a metal cylinder with spokes in it, basically a large version of the spoked metal cylinder used in the music boxes that were very popular among the well-to-do during de Vaucanson's time and can sometimes still be seen in antique shops today.

The idea behind the de Vaucanson loom was ingenious and technically sound. Prototypes of the de Vaucanson loom worked reasonably well. A big problem, though, was that the metal cylinders were expensive and difficult to make. Moreover, by their very nature they could only be used for making images that involved regularly repeated designs. True, by switching to new cylinders it would be possible to produce designs of open-ended variety, but in practice the constant switching over of cylinders proved too time-consuming and laborious. A few examples of the de Vaucanson loom did go into production, but it never really caught on and was soon discontinued.

By the late eighteenth century, the entire wealth and might of the Lyons silk-weaving industry—by far the largest silk-weaving industry in the world—was stymied by the fact that Lyons weavers did not have access to an efficient loom. They yearned for a machine that would allow a great deal more silk fabric to be woven in a day than the maximum one inch that could be produced by a weaver and draw-boy working flat out.

The enforced lethargy of the rate of production kept Lyons weavers poor. Even the master-weavers who headed weaving studios, and the merchants who sold the fabric to wealthy customers, suffered hard times. The world craved a flood of Lyons silk, and all Lyons could offer was a slow trickle.

·3·

The son of a master-weaver

The invention of Jacquard has produced a general and total revolution in the procedures for manufacturing; it has traced a great line between the past and the future; it has initiated a new era in general progress.

The Count of Fortis,
Éloge historique de Jacquard, 1840

The wealth of Lyons today is founded on the city's high-tech, banking, construction, chemicals, food, and printing industries. Many of the major Lyons corporations in these industries are headquartered in the district known as Part-Dieu, which is located to the east of the Rhône. The head offices often occupy some of the boldest and most original modern architecture in France. Part-Dieu even has its own skyscraper, the 165-metre (541-feet) Credit Lyonnais tower. Light-brown in colour, the tower looks like a giant cigar, as if symbolizing the lifestyles of conspicuous consumption to which the bank's customers might hope to aspire.

But not all districts of Lyons denote conspicuous consumption. If you leave the prosperity of the modern business area of

Part-Dieu behind you, cross the Rhône by bridge or ferry, and head for the district of Croix Rousse, you will encounter a very different atmosphere.

The name Croix Rousse means 'russet cross' and derives from a large cross of local russet-coloured stone that could indeed be found at the very highest point of the district from the Middle Ages to the time of the French Revolution. For many centuries a mere backwater in Lyons, Croix Rousse rose to prominence in the early years of the nineteenth century, when the abundance of space it offered close to the city centre encouraged Lyons's burgeoning weaving industry to move to the district. By the 1830s, Croix Rousse was the heart of the world's silk-weaving business.

But the days when Croix Rousse was a riot of bustle, haste, and activity are no more. Today, the pavements no longer reverberate with the endless noisy rattle of hand-looms being operated by the more than 30 000 weavers who worked in Croix Rousse in its heyday. No brawny workmen load heavy wrapped silk fabrics into the carefully cleaned backs of horse-drawn carts. No fat, prosperous merchants linger over their glasses of absinthe at roadside cafés, boasting to their friends of the latest deal they made with some wealthy customer. Instead, today, there is mostly silence, the occasional bark of a dog, sometimes a Renault or Citroën dawdling along, and here and there vaguely discontented couples walking arm-in-arm, out for an evening in picturesque Croix Rousse before heading back to their modern apartments on the other side of the river. They look rather demoralized, as if they find the district a little too quiet for their liking.

After its glorious nineteenth century, Croix Rousse has long sunk into obscurity. Yet despite this, silk-weaving has not entirely vanished from the district, for a faithful band of about two dozen expert weavers still weave silk there by hand. I visited one of them, Georges Mattelon, in his studio. Cluttered with three Jacquard hand-looms and a myriad bobbins of coloured silk thread, the studio smelt of old wood and oil and was as comfortable as a hobbit-hole. Georges, a weaver in Croix Rousse for six decades, showed me the very loom on which he wove much of Queen Elizabeth's dress for her wedding in 1947. Later, he took me for lunch at a nearby *bouchon*—the local name for a bistro—where six or seven other hand-loom weavers were gathered. Over steak, *frites*, and red wine, this band of silk-weaving old-timers spoke with enormous enthusiasm of the glory days of the Lyons weaving industry in the nineteenth century. It was as if they might have been working as weavers then, too.

Strolling around Croix Rousse today, you cannot escape the feeling of walking through a ghost town. There are restaurants and bars, but none of them is very busy. The general impression is of a rundown, fairly shabby district which does at least have considerable picturesque charm. Its narrow alleys wind among pretty, robust-looking buildings whose plaster walls are painted in light pastel hues. All the main streets lead to *La Place de la Croix Rousse*—the district's main square. The square's bars and bistros, well off the usual tourist route, are patronized almost entirely by locals and serve excellent wines and delicious, modestly priced meals: Lyons delicacies such as eel stew, salted capon, *quenelles* of pike, and the rich onion soup for which the city is especially famous.

On the far side of *La Place de la Croix Rousse,* near the steep steps that lead down to the river a couple of hundred feet below, there is a large and prominent grey stone statue. The statue depicts a man in a greatcoat, with shoulder-length hair. As a work of art, it is a disappointment. Its bland, expressionless face is, indeed, the face of a statue, not a man. Its pose—the left arm vertical, clutching a rolled parchment, the right hand held rigidly across the chest—seems artificial, and the left arm is too long and out of proportion with the body.

Inscriptions on the four sides of the cubic plinth reveal some information about the man whom the statue depicts. '*To J.M. Jacquard, the benefactor of silk-workers, from a grateful city of Lyons*', reads one. Jacquard himself, remembering an enforced swim he is believed to have had in one of Lyons's waterways courtesy of irate draw-boys who had been rendered unemployed by his remarkable invention, would probably have considered this inscription more than a little ironic. *'Inventor of the loom for the manufacture of luxury fabrics'*, says the inscription on the next side of the plinth. The other inscriptions give the dates and places of Jacquard's birth and death.

While mediocre as art, the statue does at least furnish a lasting memorial to a man whose inventive labours led not only to a revolution in weaving, but also in how mankind handles information.

Joseph-Marie Jacquard was born on 7 July 1752, in the Lyons parish of St Nizier. This parish is in La Presqu'île, a few hundred yards south of Croix Rousse. Jacquard was the fifth of nine children of Jean-Charles Jacquard and his wife Antoinette. Jean-Charles was a master-weaver of brocaded fabrics. He was initially fairly prosperous.

The family lived in La Presqu'île, in a succession of apartments that had workshops attached where Jean-Charles supervised several silk-weavers. Like the sons of many Lyons weavers, young Joseph-Marie (known familiarly as Joseph) did not go to school; his father needed him to perform odd jobs in the workshop. Joseph and his sister Clémence, who was five years older than him, were the only Jacquard children to survive into adulthood. Their mother Antoinette died on 15 July 1762, when Joseph was ten years old. After her death, the family gradually slid into poverty.

Joseph worked most days in his father's workshop, growing up in an atmosphere saturated with the craft of silk-weaving. He lived surrounded by the tools of the trade: the big bobbins of dyed silk fabric, the smaller, precious bobbins of gold and silver thread and the great heavy wooden drawlooms. These, with their hundreds of warp threads and their jungle of elaborately contrived vertical cords, pulleys, and control rods, were worked from dawn to dusk to produce a few centimetres of precious, beautiful decorated fabric.

On 20 January 1765, Clémence married a family friend, Jean-Marie Barret, in the church of Saint-Nizier. Barret, a cultured man who adored books, took an interest in Joseph's education. He taught the twelve-year-old to read and told him of the greater world beyond the silk-weaving workshop. Joseph knew little of that world, but the great ships he saw on Lyons's rivers, taking the city's silk fabrics and other merchandize to customers around the world, whetted his curiosity for other places and other lives.

Barret, who became Jacquard's teacher and mentor, helped to fill in some of the gaps in Joseph's imagination. Most likely Barret spoke to the boy about politics, too. It was an age when few people outside the privileged nobility or the clergy could see much good in the French political system. Educated but landless people were particularly alive to the outrageous inequalities between those with money and those without. The rich lived lives of preposterous luxury, idleness, and gastronomic and erotic excess. The poor, for their part, dressed like scarecrows. Their diets were so desperately dependent on bread that when its price increased by a few sous they faced starvation.

In 1772, at the age of forty-eight, Jean-Charles Jacquard died. It was by no means a premature death by the standards of the day. His will revealed more assets than anyone expected: there was possibly something of the miser in his makeup. As well as the workshop and apartment in Lyons, there was a productive vineyard and even some quarries at the nearby village of Couzon-au-Mont-d'Or. In keeping with the tradition of the times, his only surviving son Joseph, now twenty years old, inherited everything.

After Jean-Charles's death, Jacquard worked half-heartedly at his father's trade, but without much success. He kept himself largely by living off his dwindling capital, never a good idea. On

26 July 1778 he married a young woman named Claudine Boichon. Their only son, Jean-Marie, was born in April 1779.

The birth of Jean-Marie evidently did not encourage Jacquard to settle down to making a success of his career. What happened to him over the next four years is not clear, but he certainly went through most of his capital with alarming rapidity. In May 1783, when he was approaching his thirty-first birthday, Jacquard confessed to Barret that he had spent almost all his inheritance. Jacquard's more romantic contemporary biographers attribute his descent into poverty to spending too much time trying to build a better loom, but in fact there is no evidence whatsoever that Jacquard was working as an inventor at this stage in his life.

A sombre chapter in Jacquard's life followed. He does not appear to have gone bankrupt, but he was certainly forced to sell almost everything that remained of his inheritance: a small house, two drawlooms, as well as his wife's modest jewellery and even—according to one source—his very own bed. He sank to the level of an unemployed workman, inevitably dragging his family down with him.

It is not known for certain what Jacquard did next. Some nineteenth-century sources have him working as a labourer in a plaster quarry, others maintain that he toiled for a lime-burner in the Bresse area. One alleges that he became a lime manufacturer, another that he worked in a gypsum mine. It is certainly known that Claudine stayed with their son in Lyons, working in a small straw-hat factory. Her wages for this, one franc a day, were barely sufficient to buy bread to keep herself and her son alive.

Jacquard was only saved from this life of poverty and obscurity by the French Revolution, which shattered and transformed France between 1787 and 1799. It is known that Jacquard helped to defend the city against the Revolutionaries during the siege of Lyons in 1793, when Lyons went through a counter-Revolutionary phase. The source of these sentiments in Lyons is not hard to identify. It stemmed substantially from the city's long commercial tradition of having royalty and aristocrats as customers.

During the battle for Lyons, Jacquard's fifteen-year-old son fought by his side. When the city fell, the two fled together. Most of the leading counter-Revolutionaries were guillotined. By the end of the siege Jacquard himself had become known as an ardent defender of Lyons against the Revolutionary forces. Had he been caught and identified he would very likely have been booked in for a morning appointment with the guillotine himself. But with a reversal of loyalties that was as dramatic as it was sensible, he and his son adopted false names and joined the Revolutionary army.

The precise nature of Jacquard's military career is in some doubt, but it is known that he progressed to an officer rank within a few months of joining. One source claims Jacquard occupied a high rank in the military government of the German city of Worms after the Revolutionaries overran it in 1797, but there is no evidence to confirm this story. Of course, if Jacquard were using a false name at the time, no evidence is likely ever to come to light. What is known is that during this period of military adventurism for Jacquard and his son, Jean-Marie was killed in a battle or skirmish, probably in Germany and most likely in 1797.

The death of his son and only child robbed Jacquard of the desire to remain a soldier. By 1798 he had returned to Lyons and to his wife. His health was far from good; he had been wounded in battle and spent several months in a Lyons hospital, ill and grieving. But he recovered, and finally managed to win his discharge from the army.

Once Jacquard left hospital he again worked at various odd jobs; repairing looms, doing occasional weaving work, bleaching straw hats, and driving light horse-drawn carts between the Lyons suburbs of Perrache and la Guillotière. It appeared that once again, after his brief military career, Jacquard had lost the plot of his life.

But the career that would catapult him into immortality was just about to start.

·4·

The Emperor's new clothes

Jacquard was a man who was most at home among workmen. He was always happiest in their company, and to know him as he really was one had to see him in his ordinary clothes in a weaver's studio, giving the weavers instruction on how to make best use of his loom.

The Count of Fortis,
Éloge historique de Jacquard, 1840

The French Revolution had given Jacquard a taste of adventure, but after the death of his son and the end of his military career, he returned to Lyons with little to his name except grief, poverty, and memories.

Today, even with the gift of hindsight, his rapid rise to fame and wealth appears surprising, almost unbelievable. Yet he was living at exactly the right time. At the start of the nineteenth century, the chaos, butchery, continually gyrating politics, and general fiasco of Revolutionary France were being metamorphosed into a coherent, ordered, disciplined new society that saw rational order as its god.

It was a society that passionately sought to be governed by reason. True enough, experience suggests that it is all too often when people are most claiming to pursue reason that they are at their most unreasonable, and no one who had lived through the Revolution and witnessed any of its horrors could have doubted this. But not even the agonizing memory of the Revolution's excesses could dull the ardour for building a new, rational society that was so deep-rooted in the French mind at the time.

This enthusiasm even went so far as to extend to something as apparently sacrosanct as the calendar. In 1793 the Revolutionary government introduced a new version. This replaced the Gregorian calendar with a supposedly more scientific and rational system designed to avoid what were regarded as superstitious Christian associations. Each new month was exactly thirty days long. All the new months were given new names. The first month of spring, for example, was known as Germinal (from the Latin word *germen*—'seed'), while the month in the height of the summer was called Thermidor (from the Greek words *therme*— 'heat' and *doron*—'gift'). The rational calendar was never particularly popular with the people—the Gregorian calendar was re-established on 1 January 1806—but its very existence shows how widespread and far-reaching the passion for rationality was in Revolutionary French society.

It was a society that had a particular love of new types of machines and for all kinds of completely new inventions. This was only to be expected; after all, machines are the ultimate physical embodiment of reason.

One nineteenth-century biographical account of Jacquard's life declares that he had already started work on his loom when the French Revolution broke out. But this, like too much that has been written about Jacquard in the past, is at best uncorroborated supposition, and at worst fantasy. The truth is that there is no evidence that Jacquard started work on loom-making before 1799: the year when Napoleon Bonaparte came to power, united France, and ended the Revolution. Jacquard had spent most of his time during the Revolution struggling to stay alive, a battle that does not allow much leisure for the calm and inspired reflection behind great inventions.

Of course, Jacquard *may* have been thinking for many years about the challenge of building a new type of silk-weaving hand-loom. For all we know this had been on his mind since his childhood days. Some of his contemporary biographers assert, plausibly enough, that he spent time as a young man working for his father as a draw-boy. If this was so he would certainly have had first-hand experience of the mind-numbing complexity and tedium of weaving pictures in silk by manual means. But biography must deal with what can be known for certain, and there is no proof available that Jacquard did, in fact, ever work as a draw-boy. Nor can the onset of Jacquard's work on *inventing* looms rather than *weaving* with them be reliably dated from any entry in any diary or workshop journal, for not a single word ever written by Jacquard survives.

Yet in place of direct evidence there is some useful circumstantial evidence available. Jacquard took out his first patent for a loom (not *the* Jacquard loom at this point) in December 1800. He had not returned to Lyons until 1798. The months he spent in hospital that year could indeed hardly have been a period when he did very much practical experimentation in loom technology, no matter how much he may have been thinking about the matter. Common sense suggests that he *must* have embarked on his new career as an inventor in 1799, or at the very latest the following year.

If indeed it wasn't until the last years of the eighteenth century that Jacquard started trying to build a better silk-weaving loom, the new political and technological mood of the country would have played a crucial role in inspiring him. This new mood had sprung up spontaneously, but had been nurtured by the personality and conviction of one man who seemed to the French to exemplify all that was best about the new society they had created: Napoleon Bonaparte.

The aspects of Napoleon's personality and vision that mattered most to Jacquard were unconnected with the new leader's political or military success. Rather, they were related to his respect and admiration for science. This, like the intensity of Napoleon's own personal ambition, lay at the core of his personality. Indeed, once Napoleon became the undisputed ruler of France in 1799 he deliberately set out to give an enormous boost to French science and industry.

The rivalry between the British and French industrial revolutions of the eighteenth and nineteenth centuries has stimulated a great deal of debate. The consensus today is that France's industrial technology was, in fact, considerably inferior to Britain's for much of the period, that it was only from the middle of the nineteenth century that France started to enjoy a dramatic level of industrial growth and expansion. At the start of the nineteenth century, the French economy was again starting to expand, but France's major area of manufacture remained the market for hand-crafted luxury goods aimed at the wealthy in Europe and America. This included items such as jewellery and—of course—silk, which was in fact France's largest export commodity in the early nineteenth century.

These observations give us a crucial opportunity to set Jacquard's work in perspective: he was trying to effect a vast improvement in a machine that played a key role in the economy of his native land because it manufactured the fabric so prized both at home and abroad. If he could succeed, any significantly improved loom would have a claim to be the most precious, and important, piece of machinery in France.

And there was even more at stake than this. A successful new silk-weaving loom would also represent one of the few instances during the late eighteenth and early nineteenth centuries when French technological innovation would have won a resounding victory over its British counterpart.

It was lucky for Jacquard in particular and for French industry generally that Napoleon had been fascinated by the silk industry all his life. When the future Emperor was a boy, his father had actually formulated plans to plant mulberries in Corsica as the basis for a Corsican silk industry and had very nearly managed to win a substantial government grant to do this. Napoleon was entirely aware of the importance of silk in the French economy, and was always ready to take steps to protect and advance the industry. On several occasions—both as First Consul and as Emperor—he visited Lyons, a city he loved, and gave speeches to inspire the city's industrial and commercial communities. He was always received in Lyons as a hero. The people of Lyons, like most of the French population, regarded Napoleon as the saviour of the nation: the great leader who had put an end to the disorder and violent chaos of the Revolution.

Even if Jacquard was indeed inspired in his work by Napoleon and the new society the Emperor was creating, making the leap from poor and emotionally traumatized artisan to hugely successful inventor must have been a struggle of epic proportions. It is not known exactly how Jacquard was supporting himself in 1799 and 1800; possibly he was again working at a variety of menial jobs. At this crucial stage of his life, and in the total absence of any reliable original sources that would satisfy modern scholars, there is little choice but to suppose that the assertions of a few nineteenth-century writers, whose work is based on hearsay, were close to the truth. This is that while Jacquard was spending his days struggling to earn enough money to buy bread for himself and Claudine, he was devoting his evenings and nights, as far as time permitted, to studying loom-making truly seriously for the first time in his life.

Jacquard must have initiated his career as an inventor with a minimum of financial and practical resources and with very little to sustain him other than the conviction that he could attain his great goal and that this goal was infinitely worth achieving. He refined his skills by making a variety of types of loom, but was

31

always aware of the loom he really wanted to build—a revolutionary new machine for weaving pictures into silk brocade. Soon he began constructing a prototype of the machine, carving the pulleys and other components himself.

Living within a community of weavers that was itself literally a tightly woven group which greatly enjoyed gossip, the story of what Jacquard was doing started to get about. Jacquard's innate talent as a craftsman and inventor made people with money think that here was a man who stood a fair chance of succeeding in a task that had been the dream of silk-weavers for more than a century. Master-weavers and merchants with cash to spare went to see Jacquard. They offered him various arrangements under which he would make his new brocade loom—assuming he could get it to work—available to them at a favourable price in exchange for immediate financial assistance upfront.

For the investors the deals must have been risky. There were, in fact, numerous would-be inventors tinkering with ideas for a mechanical drawloom. Prototypes had been developed for three or four decades, but none of them had come anywhere near fulfilling their inventors' expectations.

Most of these inventors possessed more ambition than inventive capability. Almost all of them were short of cash and desperate for patronage. But perhaps there was something particularly convincing about Jacquard's plans and approach that made him especially prone to attract support from prosperous master-weavers and dealers in silk fabric. These people were only too aware of how much their own profits would be boosted if he were successful.

If Jacquard did have patrons who invested in his labours, they did not have to wait long before seeing results. Once he started his work as an inventor, his genius for it became apparent almost at once.

The patent he took out on 23 December 1800 is registered in Lyons's municipal archives as being for a machine 'designed to replace the draw-boy in the manufacture of figured fabrics'. This loom, his first, can be seen as a kind of preliminary step on the way to achieving his grand objective. It could weave simple images into fabric by means of foot-treadles, with a limited number of treadles being used to produce small geometrical or floral designs that were highly prized for use in fabric in making men's waistcoats or women's dresses. These fabrics had been popular under the reign of King Louis XVI. They were now coming back into fashion, for an élite class was once again emerging in French society: its privileged status now based on practical achievement rather than on birth and inheritance.

Jacquard continued to work hard on developing other proto- types. He presented his treadle-loom at the second exhibition of French industry in Paris in 1801. These exhibitions had been instituted by Napoleon, further evidence of his conviction that it was as important for France to rise to pre-eminence in the physi- cal sciences and in industry as it was to be supreme on the battle- field.

Jacquard's treadle-loom was received at the exhibition with immense interest. From this point on his reputation started to grow quickly. Even for this first loom, he was rewarded with a bronze medal for technical achievement. This was handed over to him personally on 25 September 1801 by the Minister of the Interior, Jean-Antoine Chaptal. It was Jacquard's first-ever offi- cial recognition by the Government.

While continuing to work on his most ambitious loom of all, he enjoyed another diversion. In 1802, the French Society for the Furtherance of National Industry had sponsored a prize for a loom for making fishing nets. Jacquard designed a new kind of loom for this purpose. The loom was a great success, and the Society for the Furtherance of National Industry invited him to visit Paris in August 1803 at its own expense to honour him with an award for his work.

This was Jacquard's first opportunity to meet the talented mathematician, scientist, and logistics expert Lazare Carnot. Carnot's French nickname *L'Organisateur de la Victoire* ('The

Organizer of the Victory') showed the enormous esteem in which he was held. Carnot, fascinated all his life by science and new inventions, had taken a keen interest in the Society for the Furtherance of National Industry. A huge, bull-like man, he had expressed a wish to meet the inventor from Lyons who was rapidly building a reputation as a brilliant designer of looms.

Carnot's nickname derived from a terrifying moment in his career in May 1795, during the last years of the Revolution. He had found himself in a severely compromised position when an obscure deputy suddenly demanded the arrest of all the members of the Committee of Public Safety, a Revolutionary body that had been discredited and of which Carnot had been a member. 'Not guilty' was rarely a verdict to which Revolutionary courts were partial, and being arrested all too often resulted in a visit to the guillotine, and a short, sharp shock. There were a desperately tense few minutes when it seemed possible that the moment of Carnot's downfall might have come.

But Carnot was saved by another deputy springing to his defence and suddenly exclaiming 'He organized the victory!' This was a reference to the crucial role Carnot played in organizing the logistics behind the Revolutionary army's successes. General cheering broke out. That cheering saved Carnot's life. He was never in such mortal danger again.

When Carnot and Jacquard met, the great ebullient hero of the Revolution was immensely impressed with the modest, thoughtful, softly speaking inventor from Lyons. Jacquard, at fifty-one, was just ten months older than Carnot. The fact that Jacquard had fought bravely in the Revolutionary army (his counter-Revolutionary activities were either not known about or ignored) and that his son had died fighting for France, greatly endeared him to the Organizer of the Victory. Carnot presented Jacquard with the prize from the Society for his fishing-net loom. At today's prices the 1000 francs Jacquard won were worth about £20 000 ($32 000). His days of penury were over and he returned to Lyons a hero.

In 1804, the same year Napoleon crowned himself Emperor at Nôtre-Dame in Paris, Jacquard finally patented his loom for weaving brocade using punched cards to control the action of the warp threads and therefore to control every row of weaving. The success of this loom was evident from the beginning. Its effect on the French silk industry was immediate, enormous, and extraordinary.

With his enormous mechanical talent, Jacquard easily saw that the problem with the drawloom was that it did not really solve in any constructive sense the problem of how to rapidly raise and lower the warp threads to form the shed when a decorated fabric was being woven. As we have seen, the drawloom was really nothing more than an *aid* to the weaving process, just as an abacus is a helpful manual aid to calculation but not a calculating machine.

By the time Jacquard began applying himself to solving the problem, the world of machinery had advanced considerably even compared with fifty years earlier. The time was ripe for a mechanical solution to a problem that had, in a sense, bedevilled the Lyons weaving industry since its foundation.

What made the solution possible were punched cards.

The idea was this: each punched card would be pressed once against the back of an array of small, narrow, circular metal rods. Each individual rod would control the action of one weighted string that would in turn govern one individual warp thread. If the tip of the rod in question encountered solid cardboard when pressed against the card, the rod would not move and the warp thread it controlled would stay where it was. On the other hand, *if the tip of the rod in question encountered a hole in the punched card, then the tip of that rod would pass through the hole and the individual warp thread controlled by that particular rod would be raised.*

The crucial point to understand is that the precise array of raised or stationary rods (and corresponding raised or stationary warp threads) *could be different for every single line of weaving.* Put another way, every single row of weaving would have a new punched card to govern it, and of course the punched cards would all need to be processed in precisely the right sequence. But the beauty of this system was that *once all the punched cards had been made and strung together in the right sequence, that chain of punched cards would always produce the same design.*

Jacquard has gone down in history as the inventor of the punched-card loom for weaving silk brocade, but in fact it was not Jacquard who first had the idea of using punched cards in this way. Some sources attribute the original idea to a weaver with the surname Falcon, who in 1728 built a loom that did use large punched cards. It is not clear what Falcon's first name was. Some sources say Louis and others Jean. There is no way of being sure which is the right one, or perhaps he used them both. In many respects Falcon's very obscurity attests to the inadequacy of his loom.

Falcon's process was hopelessly slow and never got beyond the prototype stage. His cards were clumsily made and unreliable. But the biggest problem with the Falcon was that his loom was not *automatic.* Instead, it required the drawloom's draw-boy to press the punched card against the rods controlling the warp thread governors every time a row of stitches was required. In effect, the Falcon loom was, like the drawloom, really nothing more than a manual aid to the weaving process. It was not a *machine* at all. Jacques De Vaucanson's loom, on the other hand, *was* a machine, but an inefficient, expensive, and not very practical one.

Jacquard brought all the ideas together, refined the technology, and solved all the practical problems to create a loom that was a true world-beater. He saw that the way ahead would indeed be through the use of punched cards. These, unlike the de Vaucanson cylinders, enabled the sequence of weaving instructions to be as long as the weaver desired. The portrait of Jacquard which Charles Babbage showed at his soirées required, as we have seen, a total of 24 000 punched cards to be employed when it was being woven. In fact, the woven picture of Jacquard was an exceptionally complex image. It was not a commercial proposition but a special production to show how remarkable Jacquard's loom really was. The usual total of cards required for even the most sophisticated commercial woven fabric on a Jacquard loom was in the vicinity of 4000.

Jacquard's work illustrates an important principle of great inventions: they are rarely ideas plucked out of the air, but more often than not build on previous attempts to make a key breakthrough. The inventions that go down in history bring to those attempts a level of insight, technical mastery, and sheer genius that allows the invention to be successful. The perforated or punched card is now associated with Jacquard, not Falcon, because Jacquard made a machine that worked properly. In any case, most likely even Falcon did not originate the idea of the punched card; its origins probably lie in the obscurity of history. Invention is a harsh science, and the successful completion of a working machine or device that provides the benefits it is supposed to provide is more important than priority of concept, which may in any case be impossible to ascertain.

Besides, the really important idea—the one Jacquard certainly *did* pioneer—was the notion of *applying punched cards in the loom control system automatically* so that the loom in effect *continually feeds itself* with the information it needs to carry out the next row of weaving. It is because of this system that the Jacquard loom was a machine of a calibre and sophistication that had never been seen before. In fact, when it was patented in 1804, it was unquestionably the most complex mechanism in the world.

Jacquard built an *automatic* loom that was reliable, rapid, and which combined all his ideas into an effective, commercially viable, and mechanically stunning contraption. Using it, the weaver did not need an assistant at all. The weaver alone had the ability to control the mechanism that manoeuvred all the warp threads into the position they needed to be in for each successive row of weaving. The weaver alone would also handle the shuttle that would be passed through the shed once the shed was prepared.

Above all, the Jacquard loom was infinitely *flexible*. Any image or decoration could be embodied into the chain of punched cards and woven by the Jacquard loom. Jacquard could have used his loom to weave an image of a laptop computer, had anybody known how to draw one. If they had, they would have been drawing one of the loom's descendants. In any event, the Jacquard loom had been born.

And how *rapidly* did it weave? The astonishing truth is that the Jacquard loom enabled decorated fabric to be woven about *twenty-four* times more quickly than the drawloom. Whereas in the past even the most skilled weaver and draw-boy duo could only manage two rows or picks of woven fabric every minute, a skilled lone weaver using the Jacquard loom could manage to fit in an average of about forty-eight picks per minute of working time.

This was a prodigious gain in speed for the technology of the time. We can more readily appreciate the impact of the speed increase when we consider that, today, a supersonic jet aircraft flies at up to about twenty-four times the average speed of a motor car. The increase in speed was as remarkable as that.

Using the Jacquard loom, it was possible for a skilled weaver to produce two feet of stunningly beautiful decorated silk fabric *every day* compared with the one inch of fabric per day that was the best that could be managed with the drawloom.

The loom also took all the hard work and difficulty out of making the image. Once the cards had been made for a particular design, the weaver merely had to operate the shuttle and advance the chain of cards, one at a time, usually by means of a foot-treadle. For the first time in the history of silk-weaving the fabulously ornate designs—works of art in themselves—could be produced automatically. A full technical account of how the Jacquard loom works can be found in Appendix 3.

Today, some of the fabrics produced on a Jacquard loom in the first decades of the eighteenth century are on display at the Museum of Weaving and Decorative Arts in Lyons. The colours of the fabrics are slightly faded, but the beauty of the images remains stunning, warning us never to underestimate the technology of the past.

After Carnot, the next French VIP to acknowledge Jacquard's achievement was Napoleon himself. Napoleon had never beaten the British militarily, but here was a French machine that had won a massive, comprehensive, advantage over British technology. The small man was delighted. And the fabrics it wove also pleased Napoleon greatly. They were rich, lustrous, and beautiful: not unlike how he perceived himself, in fact. The days of the pampered French nobility were over, but luxury never goes out of fashion.

During Napoleon's reign as Emperor from December 1804 to April 1814, he insisted that his ceremonial clothes be woven by the silk-weavers of Lyons. From 1805 onwards, this meant the Emperor's clothes were invariably woven on a Jacquard loom.

Napoleon and Jacquard met for the first time on 18 April 1805, when the Emperor and his wife Josephine visited Jacquard's Lyons workshop. Three days earlier, on 15 April 1805, Napoleon had issued a decree declaring the Jacquard loom public property. The same decree compensated Jacquard by awarding him a handsome annual pension for life of 3000 francs (about £60 000 or $96 000 in modern terms) along with a royalty of 50 francs (about £1000 or $1600) for every Jacquard loom brought into use in France, with the royalty applying for a period of six years. This generosity on the part of the French Government towards a successful inventor had been official policy even before the Revolution. Several other inventors of different kinds of looms had been awarded government grants during the eighteenth century, but none of their inventions remotely rivalled Jacquard's in stature or importance.

During his meeting with Napoleon, Jacquard expressed, in his modest and quiet way, a keen gratitude for the pension and the royalty his Emperor had granted him. A delighted Napoleon, hearing this, turned towards his entourage of courtiers, advisers, and bodyguards. 'Think of all those who incessantly come to me with demands for government grants and other financial favours!' he exclaimed, 'and here is a man of vast talent and industry who is happy with so little!' Napoleon took to Jacquard from the first, and played an enthusiastic personal role in encouraging the use of the loom throughout France.

After this recognition at the very highest level, Jacquard basked in prosperity and success. In 1819, when he was sixty-seven years old, the Government awarded him a gold medal, as well as the highly coveted Cross of the Legion of Honour, for his original 1804 invention of the automatic punched-card loom. By the 1820s, after years of helping the weavers of Lyons to make the most of their Jacquard looms, Jacquard accepted that the loom was now as efficient as he could make it with the prevailing technology. He retired to the pretty village of Oullins, a few miles from Lyons, and now a suburb of the city. There, Jacquard enjoyed a prosperous rural life, even herding sheep on land he had purchased. He died peacefully at Oullins on 7 August 1834. Claudine had pre-deceased him, in 1825. This reputation went on growing after his death. An engraving of 1841 shows an important visitor to a Lyons weaving studio receiving a woven portrait of Jacquard.

There is a story, still spoken of in the few remaining weaving workshops of Croix Rousse, that Jacquard was once accosted by angry draw-boys. Many of them were, in fact, grown men rather than boys. They were furious with him for having invented a machine that had put them out of work. They pushed him around, and ended up throwing him into the Rhône or the Saône. Perhaps it really happened, perhaps it did not. It is certainly a dramatic enough illustration of the fact that brilliant new tech-nology does unfortunately tend to throw people out of work. But the path to acceptance of a new invention in the world at large is rarely a smooth one.

Visite de Mᵍʳ le Duc d'Aumale a la Croix-Rousse, dans l'atelier de M. Carquillat.
le 21 Aout 1841

A visit by the Duke D'Aumale in 1841 to the Croix Rousse studio of the master weaver M. Carquillat. This image, like the Jacquard portrait, is astonishingly enough a *woven* picture, not an engraving. It depicts the Duke receiving a copy of the woven Jacquard portrait (Lyons Museum of Textiles)

The sheer popularity of the Jacquard loom over the decades that followed testifies to its effectiveness. By 1812 there were about 11 000 Jacquard looms in use in France, and by 1832 there were already about 600 Jacquard looms in Britain, despite energetic French efforts to keep the technology of the Jacquard loom secret. It would, after all, be from Britain, the cradle of the Industrial Revolution, that the use of the Jacquard loom would spread worldwide, rather than from France. In Britain too, in the 1830s, the steam engine was first used to power the loom by means of a drive-belt. The powered loom grew rapidly in popularity. And with the dawn of the twentieth century, electricity started to be employed as a more precise and flexible power source.

Yet there have always been Jacquard weavers who adopted a purist approach, especially when weaving the most meticulous and valuable fabrics. As we have seen, there are still Jacquard weavers in Lyons today who operate their looms by hand. They produce luxurious brocade fabrics for business tycoons, stars of the movie and music businesses and other wealthy people. Most of Europe's royalty have been deposed, and those who remain usually have to be on guard against being seen to live too extravagantly. The new kings and queens of show-business, fashion, and industry are, in a sense, the descendants of the European royalty who first created the demand for decorated silk fabric that led to the prodigious growth of Lyons as the centre of the world's silk-weaving industry.

Today, modern fully automatic and powered Jacquard looms continue to embody leading-edge technology. This enables them to operate at speeds undreamed of by Jacquard himself, yet these modern Jacquard looms make use of exactly the same principle embodied in the loom Jacquard patented in 1804. This is tacitly acknowledged in the fact that these modern looms are also known as Jacquard looms, even today. And within the modern clothing industry, the name Jacquard has achieved the accolade of being written with an initial lower-case letter to denote 'jacquard' fabric: intricately decorated material of silk or other fine fabrics woven on modern Jacquard looms.

These modern Jacquard looms make use of computerized image scanners. These allow any visual image that is to be woven to be inputted into the loom. The scanner, in turn, is linked to a computer that converts the image into pixels in a computer program. This program is used to control the hooks that lift the correct warp threads to form the image during the weaving process.

The speed of the modern Jacquard looms, and the intricacy of the fabrics they can weave, are alike dazzling. Modern Jacquards weave familiar types of fabric such as clothing and home furnishings, as well as fabrics for unfamiliar, even surprising, purposes. For example, some modern Jacquards weave the tubes of synthetic fabric employed in artificial valves implanted inside heart bypass patients. Here, the weaving takes place under special sterile conditions. Air bags employed in cars are also usually woven on Jacquard looms. In these cases, and in others where the fabric must meet extremely demanding specifications, a Jacquard loom is used because of the strength, flexibility, and utter precision of the fabrics it weaves.

The story of Jacquard's idea might easily have ended with the new kind of loom transforming the French silk industry and then the world's. But once the idea had been born, it was taken up by inventors as inspired and brilliant as Jacquard himself; inventors who were fortunate enough to have access to much better technology than he did. Jacquard had applied his revolutionary automation idea to the industrial activity he knew best. But the potential of the Jacquard loom extended far beyond silk-weaving.

In England in 1836, Charles Babbage never had the slightest doubt about *that*. It is his inspired use of Jacquard's idea, incorporated into what is considered—with justification—to be a Victorian computer, that takes our story into a new country and—in truth—into a new dimension.

·5·

From weaving to computing

```
1 1 1 1 ▌ 1 1 1 1 1 1 1 1 1 1 1 1 1 1 1 1 1 1 1 1 1 1 1 1 1 1 1 1 1 1 1 1 1 1 1 1
2 2 2 2 2 2 2 2 2 2 2 2 2 2 2 2 2 2 2 2 2 2 2 2 ▌ 2 2 2 2 2 2 2 2 2 2 2 2 2 2
3 3 3 3 3 3 3 3 3 3 3 3 3 3 3 ▌ 3 3 3 3 3 3 3 3 3 3 3 3 3 3 3 3 3 3 3 3 3 3
4 4 4 ▌ 4 4 4 4 4 4 4 4 4 4 4 4 4 4 4 4 4 4 4 4 4 4 4 4 4 4 4 4 4 4 4 4 4 4 4
5 ▌ ▌ 5 5 5 5 5 5 5 5 5 5 5 5 5 5 5 5 5 5 5 5 5 5 5 5 5 5 5 5 5 5 5 5 5 5 5
6 6 6 6 6 6 6 6 6 6 6 6 6 6 6 6 6 6 ▌ 6 6 6 6 6 6 6 6 6 6 6 6 6 6 6 6 6 6 6
7 7 7 7 7 7 7 ▌ 7 7 7 7 7 7 7 7 7 7 7 7 7 7 7 7 7 7 7 7 7 7 7 7 7 7 7 7 7 7
8 8 8 8 8 8 8 8 8 8 8 8 8 8 8 8 8 8 8 8 8 8 8 8 8 8 8 8 8 8 8 8 8 ▌ 8 8 8
9 9 9 9 9 9 9 9 9 9 9 9 9 9 9 9 9 9 9 9 9 9 9 9 9 9 ▌ 9 9 9 9 9 9 9 9 9 9
0 0 0 0 0 0 0 0 0 0 0 0 0 0 0 0 0 0 0 0 0 0 0 ▌ 0 0 0 ▌ 0 0 0 0 0 0 0 0 0
```

I will tell you the events of the last few days. I am married and have quarrelled with my father. He has no rational reason whatever; he has not one objection to my wife in any respect. But he hates the abstract idea of marriage and is uncommonly fond of money.

Charles Babbage, in a letter to John Herschel, 1 August 1814

A winter day in London, December 1839.

It has not been a good year. With riots erupting almost every day in the countryside and towns over high food prices and low wages, many fear that England is in danger of sliding towards anarchy. By modern standards the vast majority of people in Britain are horribly poor, suffering routinely from malnutrition, illness, and despair. The small proportion of the population who are well-fed and privileged sleep uneasily in their comfortable beds, only too aware what happened in France a few decades ago.

Jacquard, resting in peace in his grave at Oullins, has been dead for five years. Queen Victoria, still only twenty, has reigned in England since June 1837.

And now here we are at Number One Dorset Street, Charles Babbage's London home. Babbage, forty-seven years old, is sitting at his writing-desk in his study. He takes out his pen, dips it into an inkwell, and starts to write a letter to a Parisian friend.

The particular friend Babbage is writing to as we meet him is the French astronomer and scientist, François Jean Dominique Arago. Babbage got to know Arago in Paris back in 1819 when he travelled there with John Herschel, a close companion from his Cambridge days. Babbage and Arago hit it off at once. They have been friends ever since. When Babbage corresponds with Arago he does so in English, while Arago replies in French. They both understand each other's native languages, but prefer to express themselves in their own.

'My dear sir,' Babbage writes:

> I am going to ask you to do me a favour.
>
> There has arrived lately in London … a work which does the highest credit to the arts of your country. It is a piece of silk in which is woven by means of the Jacard [sic] loom a portrait of M. Jacard sitting in his workshop. It was executed in Lyons as a tribute to the memory of the discoverer of a most admirable contrivance which at once gave an almost boundless extent to the art of weaving.
>
> It is not probable that that copy will be seen as much as it deserves and my first request is, *if* it can be purchased, that you will do me the favour to procure for me two copies and send them to Mr Henry Bulwer at the English Embassy who will forward them. If, as I fear, this beautiful production is not sold, then I rely on your friendship to procure for me *one* copy by representing in the proper quarter the circumstance which makes me anxious to possess it.

This letter is the first mention in Babbage's surviving correspondence of the woven portrait of Jacquard—the exhibit which, the following year, he was to make a conversation piece at his Saturday soirées. Babbage was so fascinated with Jacquard and the 'most admirable contrivance' he had invented that he asked Arago, in the same letter, to send 'any memoir which may be published of M. Jacard'. Money was no object to Babbage, so keen was he to get what he wanted. Although he was mis-spelling Jacquard's name, he had no misapprehension about the revolution the Jacquard loom had created in the story of technology:

> Whatever these things may cost, if you will mention to me the name of your banker in Paris I will gladly pay the amount into his hands and shall still be indebted to you for procuring for me objects of very great interest.

Babbage's letter then proceeds to the hub of the matter. The Englishman explains exactly why he is so fascinated by the Frenchman's work:

> You are aware that the system of cards which Jacard invented are the *means* by which we can communicate to a very ordinary loom orders to weave *any* pattern that may be desired. Availing myself of the same beautiful invention I have by similar means communicated to my Calculating Engine orders to calculate *any* formula however complicated. But I have also advanced one stage further and without making *all* the cards, I have communicated through the same means orders to follow certain *laws* in the use of those cards and thus the Calculating Engine can solve any equations, eliminate between any number of variables and perform the highest operations of analysis.

Among Charles Babbage's many contributions to the birth of information technology, the most significant was that he spotted a way to adapt Jacquard's punched-card programming to a completely new purpose: *mathematical calculation*.

At a technical level, Babbage really did borrow Jacquard's idea lock, stock, and barrel. Babbage saw that just as Jacquard's loom employed punched cards to control the action of small, narrow, circular metal rods which in turn governed the action of individual warp threads, *he himself could use the same principle to control the positions of small, narrow, circular metal rods that would govern the settings of cogwheels carrying out various functions in his calculating machine.*

As a result of this insight, Babbage was able to design the only machine of the entire nineteenth century that was even more complex than Jacquard's loom.

The conceptual link Babbage made between his own work and Jacquard's is beyond doubt one of the greatest intellectual breakthroughs in the history of human thought. It is a leap of the scientific imagination that is too easy to take for granted today, when computers and information technology are all around us, when we are so familiar with the role that computers and the Internet play in our lives, and when we use on a routine, daily basis, machines that are essentially special kinds of Jacquard looms built to weave information rather than fabric.

This is not to imply that computers would never have come about if the Jacquard loom had never existed. Computers are so useful that it is difficult to believe a technologically sophisticated society would not have invented some other machines for processing information if there had never been a Jacquard loom. But if the Jacquard loom *had* never existed, computers would certainly look and work very differently from how they look and work today.

Babbage's Difference Engine, with which he was occupied for around twelve years, from 1821 to about 1833, was to be an automatic cogwheel-based machine designed to calculate and print mathematical tables. Brilliantly ingenious, as it was, it still fell far short of the fully automated and versatile general mathematical machine—in effect a Victorian computer made from cogwheels—that Babbage eventually glimpsed on the most distant regions of his intellectual horizon. The precise link with Jacquard's work does not appear to have occurred to Babbage until the mid-1830s, when he conceived of a much more ambitious and complex device than the Difference Engine. He later christened this new machine the Analytical Engine.

Charles Babbage at the time when he was working on the Difference Engine.

Babbage planned the programming system of the Analytical Engine to be what was essentially a direct imitation of Jacquard's method for using punched cards to 'program' his loom. Babbage borrowed Jacquard's method to control every single aspect of the Analytical Engine's operation, with the cards intended to be used to input data and to make his cogwheel computer carry out specified functions. Like modern computers, the Analytical Engine was designed to have a memory (Babbage called this the 'store') and a processor (the 'mill', as Babbage named it). Babbage, a perfect gentleman when it came to acknowledging his sources, freely admitted his debt to Jacquard. He was only one of Jacquard's disciples, but he was the most brilliant, the most industrious, and the one who was least willing to take no for an answer.

The man whose mind sparked a flaming suspension bridge between the weaving industry and computing was born in London on 26 December 1791. His father, Benjamin Babbage, was a wealthy goldsmith and banker. The two professions were then closely linked; it was a small step for customers who were buying gold and jewellery from a goldsmith to use the safes in the goldsmith's offices to store all their valuables.

The Babbages had been well-established since the late seventeenth century in Totnes, a small town in the county of Devon in the south-west of England. The very first documentary mention of the Babbage family there dates from 12 November 1600, when a William Babbage, who died in 1633, is recorded as having married one Elinora Ashellaye. Later, on 18 April 1628, a Roger 'Babbidge' is listed as a payer of rates. Benjamin was the latest in a long line of Babbages who distinguished themselves in commerce. Charles Babbage's grandfather, also a successful goldsmith and also called Benjamin, had been mayor of Totnes in 1754.

Totnes today is a busy, friendly little market town, especially popular with people living New Age and alternative lifestyles. Its population has remained unchanged at about 6000 since the end of the eighteenth century, when it was extremely wealthy by the standards of the day. The historical prosperity of Totnes derived from wool sheared from the backs of the innumerable sheep that spent their lives munching the grass of the meadows on the hills that ripple around the town. This wool was knitted by hand into an inexpensive, coarse, long-lasting woollen cloth called kersey. There was a huge demand for this cloth throughout Britain and abroad for workmen's breeches and trousers.

Benjamin Babbage gradually built up his activities in the town and the surrounding district. He did not open a bank, but traded more informally, lending out sums, transacting business under his own name, and acting as an agent for some London banks. Business was excellent.

Yet by the start of the 1790s, the Totnes cloth trade was visibly waning. Machines powered by steam were making an impact on weaving and on all aspects of fabric-making. The new form of power provided what seemed at the time to be close to unlimited energy. The steam engine also offered the enormous advantage that it was no longer necessary for mills and factories to be located near running water for operating water-wheels. Coal was the fuel of the future, and in this new industrial world Totnes was at a grave disadvantage, for Devon had no coal at all. The Industrial Revolution was gathering momentum. Totnes was being left behind.

A wily fellow by nature, Benjamin was quick to spot the significance of the new developments. He rapidly made plans to transfer his business activities to London, a radical move indeed in those days, when the vast majority of the population lived out their lives in the village or town where they had been born. It has been said that many people who lived in villages at this time, and for another century or so, never met more than about seventy-five people in their entire lives.

Benjamin moved to London in 1791, taking with him his wife, Betty, whom he had married the year before. He had first-class business contacts in the capital. Benjamin eventually became a partner of Praeds Bank in London, probably one of the banks for which he had acted on an agency basis back in Devon.

His timing for his move to London was highly fortuitous, yet Benjamin had always made his own luck, and there was really nothing accidental about his success in London. He chose to migrate to the great capital—then easily the largest city in the world—at a time when there was a huge increase in the demand for credit, mainly caused by the burgeoning Industrial Revolution. The banking business was literally a golden opportunity for lenders who could keep their heads and who had the skill to distinguish good credit risks from bad. Benjamin possessed that skill. As the novels of the time were wont to put it—he 'prospered exceedingly'.

The only surviving portrait of Benjamin shows a man with a rather jovial expression, and the look of having a precise under-

Benjamin Babbage: a harsh and money-obsessed
man whose money funded his son's life's work.

standing of his importance in the world and his success. Little
else, however, is known about his personality except what can be
inferred from the letters about him written by his son Charles
and Charles's wife Georgiana.

Judging from the letters, Benjamin was frequently prone to
moods that were anything but jovial, at least in how he treated
his eldest son. Indeed, the verbal picture these letters paint of
Benjamin suggest that he could on occasion resoundingly con-
form to the now almost clichéd image of the wealthy, nineteenth-
century businessman, the sort of man for whom money has
become not so much a simple measure of business success as an
all-consuming religion. Benjamin was often impatient with
Charles and abusive to him, frequently accusing him of failing to

make serious career plans for the future, even indeed initially refusing to approve his wish to marry Georgiana until he had made safe headway in some suitable recognized profession. Charles found his father's attitude difficult to fathom; Georgiana came from a family of quality, had a fortune of her own, and was by all accounts a thoroughly charming and good person. But Benjamin believed that young men should make money a higher priority than matrimony.

The truth, however, was that Charles *had* chosen a profession, but that unfortunately it would not be seen as one until about 1950.

Yet no matter how testy and troublesome Benjamin may have been, scientific history owes him a debt of gratitude, for his preoccupation with money was to fund his son's life's work. On Benjamin's death in 1827, Charles inherited almost his entire fortune. The legacy, including Benjamin's cash in the bank and his silver and gold plate, was worth about £100 000.

To set this amount in perspective, when Charles Dickens died in 1870 after a lifetime of working harder than almost any writer has ever worked, he left about £98 000 in his will. Between 1827 and 1870 there was almost no change in the value of money in Britain, so Charles Babbage inherited more than the fruit of Dickens's life's work. Furthermore, Babbage was still young and healthy enough to enjoy it.

How much would Babbage's £100 000 be worth today? A reasonable rule of thumb is that during the first seventy years of the nineteenth century, amounts of money should be multiplied by about fifty times to give an *approximate* idea of what they would be worth today. Taking everything into account, the legacy which Benjamin left to Charles was worth about £5 million (about $8 million) in today's money. It freed Charles Babbage from financial care for the rest of his life and made possible the liberation of his scientific imagination.

There is a particular significance in how Charles's scientific career was funded. So many of Benjamin's customers had made a fortune from the cloth industry that it is really no exaggeration to say that Charles Babbage's life's work was itself financed by the cloth business. This being so, there was a sense in which when he made such creative use of Jacquard's ideas in his plans for the Analytical Engine, he was coming home.

As a boy growing up in a wealthy London family, Charles Babbage showed signs at a very early age of a fascination with engineering and mechanics. In his autobiography *Passages from the Life of a Philosopher* (the word 'scientist' was not current until later in the nineteenth century) he writes:

> From my earliest years I had a great desire to enquire into the causes of all those little things and events which astonish the childish mind. At a later period I commenced the still more important enquiry into those laws of thought and those aids which assist the human mind in passing from our received knowledge to that other knowledge then unknown to our race. Truth only has been the object of my search, and I am not conscious of ever having turned aside in my enquiries from any fear of the conclusions to which they might lead.

One particular childhood memory describes how he loved to take things apart to find out how they worked:

> My invariable question on receiving any new toy, was 'Mamma, what is inside of it?' Until this information was obtained those around me had no repose, and the toy itself, I have been told, was generally broken open if the answer did not satisfy my own little ideas of the 'fitness of things'.

The first reference to machinery in *Passages* is a reminiscence that opens Babbage's chapter about his boyhood. When he was living in London with his parents, his mother took him to several exhibitions of machinery, including one in Hanover Square, organized by a man who called himself 'Merlin'.

I was so greatly interested in it, that the exhibitor remarked the circumstance, and after explaining some of the objects to which the public had access, proposed to my mother to take me up to his workshop, where I should see still more wonderful automata. We accordingly ascended to the attic. There were two uncovered female figures of silver, about twelve inches high.

One of these walked or rather glided along a space of about four feet, when she turned round and went back to her original place. She used an eye-glass occasionally, and bowed frequently, as if recognising her acquaintances. The motions of her limbs were singularly graceful.

The other silver figure was an admirable *danseuse*, with a bird on the forefinger of her right hand, which wagged its tail, flapped its wings, and opened its beak. This lady attitudinised in a most fascinating manner. Her eyes were full of imagination, and irresistible.

These silver figures were the chef d'oeuvres of the artist: they had cost him years of unwearied labour, and were not even then finished.

The sombre note with which Babbage ends what started as a pleasant boyhood recollection carries a piercing poignancy in relation to the struggle that dominated his own life.

Charles Babbage's formal education was rather haphazard. He attended a small school near Exeter, and then another close to Enfield, in those days a small town north of London. Later he studied with a clergyman-tutor in Cambridge before returning to Devon to be educated at Totnes Grammar School. The final step in his pre-university education was a period spent studying Classics under an Oxford tutor, also in Totnes. Despite this varied schooling, by the time Babbage went up to Trinity College, Cambridge, in 1810 he was able to deal with mathematical questions in the three different types of mathematical notation

then current: the ones used by, respectively, the great mathematicians Newton, Leibniz, and Lagrange.

Cambridge—like Oxford—had an illustrious reputation for
learning, but the teaching of mathematics there left a great deal
to be desired. Maths was not seen as a practical research topic
that might have a key role to play in the world. Instead, it was
regarded as a mental training for future clergyman, lawyers, and
gentlemen who would live lives of genteel but unspectacular, and
for the most part unproductive, learning. The emphasis was on
intensive coaching to train students to solve problems that would
be set in the examinations held at the end of the year (in practice,
many examinations actually took place in January). These problems were frequently extremely difficult, but they did not have
any intrinsic value other than to hone mental agility. They were
basically a kind of mathematical crossword puzzle: ingenious,
arduous, mentally stimulating, but ultimately a trifle pointless.
Here is an example. It must surely rank as one of the most tortuous questions ever inflicted on an examination candidate.

A question from the Cambridge Tripos examination, Monday, 17 January 1814

This problem was set for Cambridge undergraduates taking their
final examinations in the 'Tripos': the mathematics examination
that would lead to a Bachelor of Arts (BA) degree being awarded
to successful candidates. Babbage himself probably tackled this
very question, for he graduated with a BA in 1814 and would
consequently most likely have sat the paper on which this question appeared. This daunting puzzle was only one of several
similar problems candidates were expected to answer on that
Monday morning.

> 'From the vertex of a paraboloid of given dimensions, a part
> equal to one-fourth of the whole is cut off by a plane parallel

to the base; and the frustum being then placed in a fluid with its smaller end downwards, sinks till the surface of the fluid bisects the axis which is vertical. It is required to determine the specific gravity of the paraboloid, that of the fluid in which it is immersed and the density of the atmosphere being given.'

Charles Babbage had never seen mathematics as a mere exercise in solving frightful problems like this. Instead, what *he* dreamed of doing was using the great science of quantity to achieve practical improvements in the processes that mattered in everyday life.

Certainly, the time for such an attitude to mathematics had never been riper. Britain was in the midst of an unprecedented technology revolution. Transport, communications, and above all the application of steam power to industry were giving people the opportunity to use levels of power thousands of times greater than that which the horse, or the human hand, could produce. Babbage desperately wanted to play his part in this revolution, so he withdrew from his formal curriculum at Cambridge and pursued his own mathematical and scientific agenda. At the time, gentlemen scholars were allowed to do this.

With two friends he had met at Cambridge—John Herschel (son of the famous astronomer Sir William Herschel) and one George Peacock—Babbage formed what he called the 'Analytical Society'. Its main objective was to overhaul the study of calculus at Cambridge and replace the notation of Newton with what Babbage and his friends regarded as the much more efficient notation invented by Leibniz. The campaign was, in the end, successful, although it would not be won until after Babbage graduated from Cambridge in 1814. But the vigour of the arguments put forward to support the change forced the outside mathematical world to start to take notice of the founders of the Analytical Society, and particularly of Babbage and Herschel.

The friendship between these two was the first serious intellectual friendship either had had. They were touchingly good companions. They addressed each other as 'Dear Herschel' or 'Dear Babbage' in letters: an extremely intimate salutation by the formal standards of the time. The informality of their letters (which usually contained abundant mathematical formulae as well as more personal material) is neatly explained by a comment Herschel made at the start of a letter he wrote to Babbage on 25 February 1813:

> When men with common pursuits in which they are deeply interested, correspond on the subject of those pursuits, the trifling ceremonials of an ordinary correspondence may in great measure be waived.

Babbage also enjoyed the social opportunities Cambridge offered. He played chess and whist and helped to form other, more jovial societies which looked into matters of comic interest to their members. One was a Ghost Club, where Babbage and his friends earnestly discussed the supernatural. Another was the so-called 'Extractors Club'. This made plans to extract any member from the madhouse should his relatives ever manage to get him sent to one. The disadvantage, perhaps, was that possibly the very act of becoming a member of such a club might have been a spur to relatives to do precisely what its members feared.

At the start of the summer vacation of 1811 Babbage returned to Totnes. He often spent part of the summer at a house his father still kept in the town. It was most likely during this summer that Babbage met a young woman named Georgiana Whitmore in the seaside town of Teignmouth, about twelve miles from Totnes. When Babbage first got to know Georgiana he was nineteen, she a year younger. She had fine, delicate features and beautiful golden-brown hair. All the evidence suggests that Georgiana and Babbage took to each other right away. The following summer they became engaged. For the first time Babbage visited her family in Dudmaston, Shropshire.

Babbage and Georgiana were married on 2 July 1814, shortly after Babbage came down from Cambridge. The couple had a comfortable income from Benjamin, who had either finally over-come his objections to the marriage, or had come to see in his practical way that the love between the two young people meant that he had been presented with something of a *fait accompli*. Babbage and Georgiana rented a house at 5 Devonshire Street, Portland Place, London. Babbage wasted no time in making his mark on the scientific scene.

During 1815 he gave a series of lectures on astronomy to the Royal Institution. In the spring of 1816 he was elected a member of the Royal Society, a learned assembly of all the great thinkers in the land. For the next few years Babbage's work was mainly mathematical. He published more than a dozen learned papers, all of which were regarded as highly competent, though not of enormous importance. In 1815 Georgiana gave birth to a son, the first of at least eight children, of whom only three survived to maturity. The Babbages led a pleasant and varied social life, with visits to friends in the country and soirées to which they invited their friends and influential people. In due course, as we have seen, the Babbages' soirées ranked among London's most famous and glittering social occasions.

It was in 1819 that Babbage and Herschel made a trip to France, the first of many excursions they made to that country to exchange information and ideas with other men of science. The two men travelled as prosperous gentlemen intellectuals. John Herschel's father Sir William was at the time one of the most famous astronomers in the world. He had taught himself astron-omy and was so enthusiastic about this science that he had devoted enormous ingenuity and effort to make telescopes for himself at home that were actually superior to the ones used at the Green-wich observatory. His discovery of Uranus in 1781 had made him famous overnight; it was the first new planet to be discovered since prehistoric times. With Sir William's reputation opening doors for the two young men, they were able to meet several prominent French scientists, mathematicians, and astronomers, Jean Arago among them.

It was very likely during this first visit to Paris that Babbage first heard of an ambitious French project undertaken at the turn of the century to make a set of reliable mathematical tables for the French Ordnance Survey. The production of the new tables, known as *Les Tables du Cadastre* (a cadastre is a register of property showing the extent, value, and ownership of land for taxation), was, at the time, the most ambitious effort of mass-calculation ever undertaken.

The project had been overseen by the eminent civil engineer Baron Gaspard Clair François Marie Riche de Prony. As his impressive name suggests, he was an aristocrat by birth. Despite often being in great jeopardy, he had managed to survive the Revolution's Reign of Terror—the period when law and justice ran amok in France, and guillotines in France worked round the clock. De Prony's survival was mainly due to certain influential Revolutionaries—chief among them Lazare Carnot himself—admiring his scientific talents.

The purpose of de Prony's undertaking was to calculate the logarithms of the numbers from 1 to 200 000 and make them available to the Ordnance Survey so that it could more easily carry out mathematical calculation. The logarithm (log) of any given quantity is the number to the power of which 10 must be raised to arrive at the quantity in question. The great advantage of logarithms is that if you know the logs of two numbers you want to multiply together, all you need to do is add the two logs together, note this total, and go back to the logarithm tables to see which number has the logarithm you end up with. That number is your answer. Eventually, tables of 'antilogarithms' were produced which simplified the business of finding the actual numeric result from the logarithmic result, but these were not available in Babbage's day.

De Prony was flattered to be asked to undertake this great project. Not surprisingly, he was terrified of failing. He urgently cast his mind around for an inspired idea that would give him a better-than-evens chance of success.

Finally, while in a second-hand bookshop, he came across a copy of Adam Smith's *The Wealth of Nations*, published in 1776. An enormously influential book, it had taught a whole generation of British industrialists about the importance of the free market and the huge increases in productivity that could result from the division of labour, with different workers specializing in different elements of the overall task.

In a famous passage, Adam Smith relates how the productivity of a pin factory he had visited had been maximized by groups of workers specializing in different stages of the production of the pins. One group of workers had, for example, straightened the wire, another cut the wire, another sharpened the tips of the pins, and so on. In this way, Adam Smith explained, the total output of the pin factory would be many times greater than that which could have been produced if each individual worker had handled every stage of the pin-making process.

De Prony decided to use the same principle to give himself the best chance of making his vast set of tables with the greatest accuracy, and within a reasonable time-frame. After planning his approach carefully, he decided to divide his human calculators into three teams.

The first team would oversee the entire undertaking. This would involve investigating and furnishing the different formulae for each function to be calculated and setting down the simple steps of the calculation process. The team would be made up of half a dozen of the best mathematicians in France, including Carnot himself and Adrien-Marie Legendre, who was famous for important work on elliptic integrals, which provided basic analytic tools for mathematical physics.

The second team of seven or eight human calculators converted the formulae into key intermediate numbers which would be the basis for the actual calculations of the values to be set down in the tables.

The third team consisted of sixty to eighty clerks whose mathematical ability was largely limited to being able to add and subtract. By virtue of the way the huge project was organized, this was all they needed to do in order to perform their necessary calculations. Curiously enough, many of the clerks were former hairdressers to the aristocracy. These hairdressers found themselves unemployed after the Revolution, for most of the aristocratic heads on which they had practised their art were no longer connected to their owners' necks. Besides, hairdressing was not immediately a top priority in post-Revolutionary France.

The tables produced by de Prony's pioneering technique occupied seventeen large folio volumes and had a reputation for being reliable. The tables impressed Babbage enormously. Their reliability was such that they were used by the French Army as late as 1940 to assist with calculations relating to surveys of terrain. But even though de Prony's tables were regarded by French mathematicians as an enormously useful asset for more than a century, they were never actually *printed*. Instead, they remained in manuscript form. De Prony arranged for the French Government to make an arrangement with a Parisian publishing company to prepare 1200 printing plates on which a substantial portion of the tables would be printed. But the printing never went ahead, apparently for cost reasons. After the huge operation to prepare the figures, the fact that the tables were never printed meant that they were not particularly useful, because only people with access to the original manuscripts could actually use them.

De Prony's mass-production approach to his enormous calculation assignment struck a chord deep within Babbage's mind. When Babbage developed his first cogwheel calculator, he decided to base his machine on a method that would reduce the extremely complex business of tables calculation to its simplest essentials, much as de Prony had. Furthermore, Babbage knew about the problems de Prony had experienced with getting his tables printed. Babbage was determined to incorporate a printing mechanism *within* his machine. This would allow the machine to produce a printed output onto paper automatically, eliminating the possibility of human error.

Babbage's plan, in fact, was that the machine itself should make printing plates that could be used as many times as required. The calculating and printing machine was to be called the Difference Engine, for reasons we explore shortly. The first great professional challenge Babbage set himself was to build it.

Meeting this challenge, and grappling with the practical and conceptual difficulties it involved, took Babbage into a realm of almost inconceivably complex and original inventiveness. He started by planning his Difference Engine; he ended by designing what was nothing less than a computer controlled by the very same type of cards that programmed the Jacquard loom.

But why was Babbage so keen to build a calculating machine in the first place?

·6·

The Difference Engine

It was not long before I made my first acquaintance with what is perhaps the most celebrated icon in the prehistory of computing, all that Babbage built of the Difference Engine No. 1—the finished portion of the unfinished engine … True to its inventor's promise, it calculates without error. It works impeccably to this day and in defiance of Babbage's detractors, provides compelling evidence for the feasibility of his early designs.

Doron Swade, *The Cogwheel Brain*, 2000

To understand why Charles Babbage chose to devote much of his life to trying to build a cogwheel calculation machine, we need to go back to an evening in the summer of 1821, when he was working in his home at Devonshire Street with his friend Herschel. The two gentleman mathematicians had been asked by the Royal Astronomical Society to help improve the accuracy of some important astronomical tables it had produced.

The Royal Astronomical Society, which still exists, admirably preserves its records of its illustrious past. In Babbage's day the

Society made extensive use of astronomical tables for the purposes of calculating the movements of heavenly bodies. Compiling these tables required many elaborate and tedious arithmetic calculations.

Babbage and Herschel were not, in fact, carrying out the calculations themselves. Instead, they were merely checking over work that had already been carried out by clerks. Curiously enough, clerks who undertook arithmetical calculations of the complex kind necessary to compile astronomical and mathematical tables were known at the time as 'computers'.

For Babbage and Herschel, even the mere checking process was extremely tedious and difficult. They were checking over the results of two independent human computers. The problem was that if they found a discrepancy between the two results they obviously had no way of knowing which was correct and so had to perform a calculation of their own. But how could they be sure that their result or one of the computers' result was right? The job really was nothing less than a mathematical nightmare.

Quite understandably, at some point during the painful labour he was sharing with his friend, Babbage suddenly exclaimed:

'My God, Herschel! How I wish these calculations could be executed by steam!'

The cry may have been a simple expression of frustration, but Babbage was not a man of empty words. Babbage was above all a practical man, and when a purpose and decisive course of action had caught fire within his mind, he possessed the energy and financial means to take steps to turn it into reality, or at least into as much of a reality as could be achieved. Later that very evening, Babbage started thinking seriously about how his objective might be attained. It was necessity that had driven Joseph-Marie Jacquard to seek to build a revolutionary type of loom. It was now necessity that was powering Charles Babbage's inventive imagination forward.

In the early nineteenth century, the lack of a reliable means of carrying out mathematical calculations was a serious problem, and the more extensive the role technology was playing in society, the more serious the problem became. As Britain's Industrial Revolution gathered momentum, the difficulty of performing complex calculations accurately became a grave limiting factor to the changes this revolution was bringing to Britain's industry and economy.

The compilation of mathematical tables such as astronomical tables, or the logarithmic tables Gaspard de Prony had produced for post-Revolutionary France, was essential for facilitating accurate calculations. But the whole problem was that *the unreliability of calculations also extended to the compilation of mathematical tables.* And of course, if you were using inaccurate logarithmic tables to carry out an important calculation, your calculation was doomed from the start.

Even worse, you had no way of knowing exactly where an inaccuracy in the mathematical tables might lie. This fact created a disturbing climate of psychological insecurity among scientists, astronomers, and mathematicians. Under the circumstances it was hardly surprising that John Herschel, writing in 1842 to the Chancellor of the Exchequer Henry Goulburn, bitterly observed:

> An undetected error in a logarithmic table is like a sunken rock
> at sea yet undiscovered, upon which it is impossible to say
> what wrecks may have taken place.

From our perspective today, with cheap electronic calculators readily at our disposal and even available on our mobile phones and watches, and with every desktop computer featuring a powerful calculator function that provides completely reliable results in less time than it takes to click a mouse, it is difficult for us to imagine *not* having access to all this calculating power. We need to make an even greater effort of imagination to empathize with the fact that early nineteenth-century scientists could not trust the mathematical tables they were using. Errors could literally appear anywhere, and there might be a terrible mistake in the midst of an otherwise perfect column or row.

A particularly painful illustration of this problem is seen in the work of the English mathematician William Shanks. In 1853 Shanks announced that he had successfully calculated to an astounding total of 530 decimal places the mathematical quantity π. This, the ratio between the circumference of a circle and its diameter, is an irrational number that begins 3.14159 and proceeds with a never-ending series of decimal places. Shanks devoted the next twenty years of his life to extending the approximation further in order to take the evaluation into new and undiscovered realms of mathematical achievement. But unfortunately for Shanks, an error in the 528th place meant that *all* his subsequent work was entirely wasted.

Babbage was not the first inventor to try to build a machine for facilitating calculation. In 1642 the French scientist and philosopher Blaise Pascal had constructed an adding machine to aid him in computations for his father's business accounts. The machine consisted of a train of number wheels whose positions could be observed through windows in the cover of a box that enclosed the mechanism. Numbers were entered by means of dial wheels. But Pascal's machine turned out to be unreliable. It never made any impact in mechanizing calculation.

The German mathematician Gottfried Wilhelm Leibniz saw Pascal's machine while on a visit to Paris and worked to develop a more advanced version. Pascal's device was only able to count, whereas the instrument Leibniz developed was designed to multiply, divide, and could even extract square roots. He completed a model of the machine in 1673 and exhibited it at the Royal Society, but Leibniz's device did not work properly either and was never anything more than a curiosity.

Pascal's and Leibniz's machines differed in one vitally important respect from the one Babbage planned. Theirs were manual devices, but Babbage was adamant from the outset that his own calculation machine would be *automatic*. Unlike Pascal's and Leibniz's apparatuses, which required meticulous human intervention every stage of the way, the machines Babbage conceived as a result of his determination to build a contrivance that would calculate tables 'by steam' were designed to produce column after column of results *themselves*, with the operator only having to turn the handle that powered the machine. It was, in other words, automatic. Even the handle-turning process could have been mechanized: there was indeed no reason why a small steam engine could not have been connected to the machine to do the job. Such a development would have answered Babbage's cry of frustration with a practical reality.

Babbage made the important practical decision about the technology at the heart of his planned calculation machine early on in his deliberations, very likely during the first week or so following his frustrated exclamation during his work with Herschel. This decision was to build his machine from cogwheels.

In the 1820s, when the science of electricity was still in its infancy and when the science of electronics, which concerns itself with the behaviour of the very particles of which electricity is composed, still lay a century in the future, Babbage had little choice but to use a mechanical technology for building his machines.

Unfortunately, he left no explicit explanation of why he decided to opt for cogwheels. But it is not difficult to guess the grounds of the decision. It is clear that cogwheels were the only technology he could imagine to be equal to the demands of the project.

Why? Because, firstly, a cogwheel is by definition a gear-wheel. It consequently facilitates the meshing with another cogwheel that Babbage saw could form the hub of the addition process at the heart of the machine he planned. Secondly, and equally importantly, a cogwheel allows the transmission of energy in regular, incremental steps. It is precisely this attribute which makes cogwheels essential in mechanical clocks. They are an essential part of the *escapement*, a device in a clock or watch that alternately checks and releases the train (i.e. the connected elements of the mechanism) by a fixed amount and transmits a periodic impulse from the spring or weight to the balance wheel or pendulum. The relationship between cogwheels and a clock's mechanism seems natural enough to us now, but like most major breakthroughs in science it took place less through inspiration than through necessity. The use of cogwheels in clock mechanisms appears to have derived from the first attempts—probably made sometime in the tenth or eleventh centuries—to build a clock that would be driven by a falling weight.

For the early clockmakers, a controlled falling weight on the end of a piece of cord was an obvious and reliable driving force to power a clock. It was, however, soon discovered that simply letting a weight fall and allowing this to turn an axis which in turn operates a clock does not actually work. The reason is that a weight accelerates as it falls. A clock driven by a simple falling weight would run faster and faster, like a clock face in a speeded-up movie. Even if a brake is used to slow the weight down, practical experience shows that the brake gradually wears away and becomes progressively smoother, causing the velocity of the weight to increase.

It is not known exactly when this problem was solved. But sometime early in the fourteenth century, an unknown genius had the idea of using a cord to connect a falling weight—whose fall was itself controlled by a speed governor—to a wheel that carried short, horizontal, uniformly spaced pegs around its circumference. These pegs meshed with an axis that was also equipped with such pegs. When the weight fell, the turning of

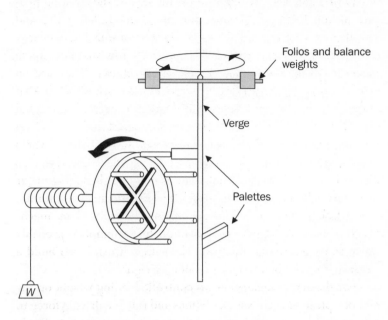

Folios and balance weights

Verge

Palettes

A pegged wheel meshing with another in an early clock.

the wheel was carried out at a regular rate because the inertia of the weight itself checked the wheel's rotation. This permitted the energy transmission to be regular and uniform, making possible the construction of clocks which work in a regular and measured way.

Eventually, it was understood that the wheel with pegs on it could be replaced by a metal cogwheel, and that this wheel would not only be more durable than a wooden one, but could also be made much smaller. As the centuries passed, the technology of clock-making became highly sophisticated. Indeed, some of the clocks of the seventeenth and eighteenth centuries are of such mechanical sophistication that they act as powerful correctives to the modern inclination to view the science and technology of the past with a kind of amused condescension.

The advent of electric and electronic timepieces during the twentieth century has tended to restrict modern clockwork mechanisms to very large clocks, or luxury watches, but some great feats of clockwork were achieved in that century none the less. For example, the Jens Olsen Astronomical Clock, displayed in Copenhagen Town Hall, Denmark and set in motion on 15 December 1955, possesses the slowest clockwork mechanism in the world. Its movements model just about every conceivable astronomical motion that one might wish to model. In particular, the so-called 'celestial pole' of the clock, mimicking the extremely slow precession of the Earth's axis in relation to the stars, will take 25 753 years to complete just one revolution.

Charles Babbage would no doubt have enjoyed the idea of such a clock. When he started work on trying to build a cogwheel calculator he was able to benefit from clock-makers' knowledge of producing cogwheels. Babbage designed his Difference Engine to use many thousands of metal cogwheels in order to perform calculations. No clock had ever employed anything like as many cogwheels as this, but Babbage was confident in his own scientific and engineering powers, and was sure he had a good chance of success.

Babbage called his first calculating machine the Difference Engine because it used a mathematical technique called the Method of Differences. This is a technique to calculate mathematical tables by repeated regular additions of the 'differences' between successive items in a mathematical series. A mathematical series is a set of terms in ordered succession, the value of each being determined by a specific relation to adjacent terms. Mathematical series can be generated by a mathematical formula. To take a simple example, the numbers 1, 2, 3, 4, 5, and so on, to infinity comprise the mathematical series requiring 1 to be added to each previous term. Mathematical tables such as logarithmic tables can also be generated by a mathematical formula.

The beauty of the Method of Differences is that it simplifies the process of calculating a long and complex mathematical series. In essence, it allows otherwise difficult multiplications to be replaced by numerous straightforward but monotonous additions. But, of course, if a machine is carrying them out, their monotony does not matter. Interestingly, Babbage's approach to calculating mathematical tables was basically the same as that chosen by Gaspard de Prony. The Frenchman also made use of the Method of Differences, and as we have seen hired unemployed hairdressers to complete the simple but numerous calculations necessary. Babbage was essentially mechanizing this process.

Babbage's mechanization was based on the idea that teeth on individual cogwheels (described as 'Figure Wheels' by Babbage) would stand for numbers, and that the machine's operation would be based around meshing independently moving Figure Wheels arranged in vertical columns with each other. This meshing process would carry out an arithmetical calculation.

Babbage decided to use the familiar, everyday counting system (base 10) for his machine. He arranged things so that the Figure Wheel at the bottom of a vertical column would represent the units, the Figure Wheel second from the bottom the tens, the Figure Wheel third from the bottom the hundreds, and so on. For example, the setting of a four-digit number such as '6538' would require the bottom Figure Wheel to be turned eight teeth to represent '8'—that is, the number of *single units*; the second Figure Wheel from the bottom to be turned three teeth to represent '3'—that is, the number of *tens*; the third Figure Wheel from the bottom to be turned five teeth to represent '5'—the number of *hundreds*; and the fourth Figure Wheel from the bottom to be turned six teeth to represent '6'—the number of *thousands*. On this occasion all the Figure Wheels above the fourth wheel would be set to zero, as they were not required.

The value of any number was therefore capable of being set in the machine in a vertical stack of cogwheels as long as there were enough wheels in the vertical stack to cover the tens, thou-

sands, tens of thousands, hundreds of thousands, millions, etc. that were required.

In his own lifetime Babbage never succeeded in completing a Difference Engine. The closest he came was building one-seventh of a Difference Engine. This one-seventh is essentially a demonstration piece. It is this portion, finished in 1832, that was the other main conversation piece at his soirées apart from the woven portrait of Jacquard.

Why exactly did Babbage fail to complete a working model of the Difference Engine during his own lifetime? To understand this, we need to understand how Babbage went about his work. With nothing but the mechanical engineering of his own age to provide him with the tools he needed, he was obliged to imagine and improvise from scratch the components he needed for his revolutionary machine. Many of the parts he designed were wonderfully complex, especially in light of the fact that nothing comparable had ever been seen before.

Babbage designed these and numerous other components without any certain initial knowledge that they would work. He took great efforts to test his designs, often making the components out of cardboard to verify their action before having them made in metal. He was a pioneer in mechanical engineering.

But unfortunately, the lack of standardization in the component manufacture at the time meant that no component could be precisely identical with the previous one, which would of course have been the ideal. It was true that parts could be manufactured nearly identically by finishing them with hand tools or by additional machining, and could subsequently be made to fit in the machine by additional filing and tweaking. But this process was labour-intensive, time-consuming, and expensive. Babbage could have benefited enormously from the existence of a precision metal industry, but there wasn't one in his day.

Babbage employed a brilliant toolmaker and draughtsman named Joseph Clement to work on the components. Clement's

(*right*) This portion of the Difference Engine, completed by Babbage and his engineer, Clement, is one of the scientific wonders of the nineteenth century.

A modern piece-part drawing of a Difference Engine component showing the remarkable complexity of Babbage's concept and the finished component built by modern precision engineering.

talents as a toolmaker were prodigious, but he was brusque and truculent. He had a habit of sending in invoices for amounts enormously greater than what had been agreed, and digging in his heels if his customers were unwilling to pay. Babbage had frequent disputes with Clement, who, like Babbage, was grappling with a task far beyond the leading-edge technology of the time. In turn, Babbage made frequent applications to his main sponsor—the British Government—for more money.

Despite these enormous difficulties, by the spring of 1822 Babbage had assembled a small working cogwheel assembly that demonstrated the operation of the Difference Engine. He announced the completion of his model in a paper to the Royal Astronomical Society dated 2 June 1822, and also in an open letter to Sir Humphry Davy, President of the Royal Society, dated 3 July 1822. Babbage deliberately wrote the letter for a wide readership. In it, he reviewed the success of his working model, described at great length the problems he had to over-come, and emphasized the expense involved. The letter received wide attention among the London scientific and mathematical worlds. It even came to the attention of Parliament. In due course, on 21 July 1823 the Treasury announced a grant of £1500 to Babbage to enable him to 'bring his invention to perfection in the manner recommended'.

Generous as the Treasury's initial grant was, the precise terms of the deal were never spelled out. In particular, the Treasury did not make it clear whether it expected the £1500 to be enough to complete the machine, or what procedure to follow if it was not. As matters turned out, it soon became clear that the £1500 was not nearly enough to allow Babbage to have any chance of completing his Difference Engine. He applied for more funds, and was granted them. Indeed, by 1834 the Government—convinced by the recommendations of influential people, and by the strength of Babbage's own conviction that the Difference Engine would be of enormous benefit to society when completed—had advanced to him a total of £17470, about £1 million at today's prices. This was a vast sum by any standards. To give some idea of how much money the Government had invested in the Difference Engine, the steam locomotive, John Bull, which was completed by Robert Stephenson and his company for shipment to America in 1831, cost £784 7s.

During this period of intense hard work, Babbage suffered two personal tragedies that stayed with him for the rest of his life. The year 1827 began with the death of his father on 27 February 1827. Benjamin, largely a self-made man, had always considered his eldest child to be a failure. By the time of Benjamin's death, the work on the Difference Engine did not add up to anything which the old man—who saw the world almost entirely in financial terms—would have considered an achievement. With Benjamin's death, Charles lost the opportunity eternally to redeem himself in his father's eyes.

A worse blow was to come. Late in August, or in the first few days of September of the same year, Babbage's wife Georgiana also died, most likely in childbirth. The precise date of Georgiana's death is not known. The only evidence that offers an approximation of a date is a letter, dated 7 September 1827 and postmarked the same day, which Babbage received from his friend, the scientist Henry Fitton, who wrote poignantly: 'I do not write with any hope that your friends can alleviate the sufferings your truly deplorable loss has caused;—for I well know the volume of what you have been deprived of—and how deeply you must feel.'

Desperate for mental distraction, Babbage went to stay with the Herschels in Slough. Babbage's state of mind remained a serious worry to his mother, who wrote to Herschel, 'I cannot expect the mind's composure will make hasty advance. His love was too strong and the dear object of it too deserving.'

About a month after Georgiana's death, Babbage left his children in the safe custody of their aunt and grandmother, and embarked on a tour of Europe. He was accompanied by one of his workmen, Richard Wright. The tour lasted more than a year, and Babbage did not return to England until late in 1828. He had no inclination to resume living in the family home at Devonshire Street. He moved to a new house at Number One Dorset Street, which was to become such a focus of London intellectual life.

Babbage continued to toil on the Difference Engine until 1834, dismayed by the lack of progress he had managed to make, yet too engrossed with the importance of the task he had set himself ever to consider giving in. Twelve years of failure had not caused this extraordinary man to waver from his course.

And then, one night in December 1834, Charles Babbage had an epiphany that radically changed not only the direction of his work and life but also the shape of the history of technology. That night he conceived an even more complex and ambitious machine than the Difference Engine, one that built on pre-industrial technology, but which left the steam age far behind.

Babbage had a vision not merely of a cogwheel calculator but of a cogwheel *computer*, and it haunted him until the end of his days.

This machine, the Analytical Engine, relied entirely on the principle behind Jacquard's loom for its operation. In the entire history of human technology, it was the first calculating machine that was capable of being *programmed*.

·7·

The Analytical Engine

The work proceeded at speed. In less than two years he [Charles Babbage] had sketched out many of the salient features of the modern computer. A crucial step was the adoption of a punched-card system derived from the Jacquard loom.

Anthony Hyman, *Charles Babbage,*
Pioneer of the Computer, 1982

Charles Babbage was thrilled about the extraordinary new horizon that had opened up in his mind. He was like an explorer who glimpses a new continent, or an undiscovered ocean, from a mountaintop.

We can get a good idea of just how excited he was from a journal entry made late on the evening of Monday, 15 December 1834 by Lady Byron, the widow of the poet Lord Byron. Lady Byron, born Annabella Milbanke, had spent the evening in Babbage's company with her nineteen-year-old daughter Ada, who was a close friend of Babbage. The distinguished self-taught lady mathematician Mary Somerville, later to found Somerville College at Oxford University, was also present.

Lady Byron, a wealthy, self-opinionated woman, was not overly fond of Babbage, but she tolerated him because of her daughter's liking for him and because of the social opportunities Babbage opened for both mother and daughter. Most of her letters that mention him do so in a disparaging way. In this particular journal entry, however, she reports on the evening with surprising neutrality, perhaps because she herself was rather caught up in his own excitement.

On this Monday evening, Babbage didn't reveal to his guests at this stage precisely *what* it was he had discovered, nor exactly *when* he had made his breakthrough. His excitement and enthusiasm, however, suggest he may have had his great intellectual epiphany not long before.

Interestingly, he spoke about his discovery in metaphorical terms rather than seeking to explain it in precise detail. Why did he choose to use metaphors? He cannot have feared that the three women would have been incapable of understanding what he was saying—they all knew his work well anyway, and Mary Somerville numbered among the leading mathematicians in Britain. Most likely what Babbage really feared was revealing too much of his discovery prematurely. Alternatively, perhaps he had not yet formulated his plans with any precision.

He told Annabella, Ada, and Mary that his first glimpse of his discovery had aroused in his mind a sensation that was something like 'throwing a bridge from the known to the unknown world'. According to Lady Byron's journal, Babbage also said that the breakthrough made him feel he was standing on a mountain peak and watching mist in a valley below start to disperse, revealing a glimpse of a river whose course he could not follow, but which he knew would be bound to leave the valley somewhere.

The closest the women got to knowing exactly *what* it was that Babbage had discovered was a remark he uttered that his stroke of insight lay 'in the highest department of mathematics'. Writing in her journal, Annabella later noted soberly, 'I understand it to include means of solving equations that hitherto had been considered unsolvable.'

In fact, Babbage's inspired idea was a plan for a completely new machine whose ingenuity, ambition, and sheer genius has deservedly gone down in history—the history of technology. And, as things turned out, Lady Byron's daughter Ada—Lord Byron's only legitimate child—would play a crucial role in communicating the importance of the new machine to the world.

The genesis of the device Babbage was to call the Analytical Engine can be traced back to a second paper on the Difference Engine that Babbage read to the Royal Astronomical Society on 13 December 1822.

In this second paper, he explained to his audience that useful as the Difference Engine was, it was always going to be handicapped by the need to *reset* the machine for each new set of calculations.

The point was that when a calculation started, the initial values that would govern it had to be entered into the Difference Engine by setting the Figure Wheels by hand. Once the Engine was properly set up, the handle could be turned to ensure that the calculation process went on automatically. In principle, the calculations would follow regularly without further intervention by whoever was operating it. But unfortunately, in some calculations, the results would start to become inaccurate as the table production progressed. The machine wasn't to blame for this. It stemmed from the fact that the calculations were based on a mathematical formula which would not in every case be *precisely* accurate for every single desired numerical result.

Eventually, Babbage found a way to design a machine that would not feature this continual slight reduction in accuracy; a machine, moreover, which could do far more than simply calculate mathematical tables. He named this machine the Analytical Engine.

Why did he call it that? This is not known for certain. Possibly it was because the machine was designed to *analyse* all kinds of practical mathematical problems and find solutions to them. As the computer historian Martin Campbell-Kelly has pointed out, Babbage's entire way of seeing the world was, in a sense, analytical: that is, he tended to solve problems by reducing them to their constituent elements and analysing those elements. In this he was in many respects typical of nineteenth-century men of science; the difference was that Babbage took analysis much further than anyone else, especially in the area of calculation.

Babbage pursued the notion of the Analytical Engine relentlessly during the months that followed his evening with the three women. Unfortunately, little is known precisely about how his work progressed. Babbage never published a comprehensive account of the Analytical Engine, let alone details of when he made all his discoveries. He made an enormous number of drawings and diagrams for the machine, and completed some small working cogwheel components designed to be used in its mechanism. But modern computer scientists, who have spent months or even years examining his plans in detail, have concluded that it would probably be impossible to build a complete working Analytical Engine from the plans without considerable additional work and without making certain assumptions about his intentions: assumptions that might well be incorrect.

None of this is to belittle the almost incredible achievement of the intellect and imagination that the Analytical Engine represents. Even though a completed Analytical Engine was never built, and most likely now never will be, Babbage's plans for the machine make clear beyond any doubt at all that the Analytical Engine would, indeed, have been nothing less than *a massive Victorian computer, made out of cogwheels.*

The machine Babbage designed would have been enormous, about the size of a small steam locomotive in his day or a large van today. It would have contained perhaps as many as 20 000 cogwheels, some mounted in vertical columns, as they were in the

Difference Engine, but others used in a variety of other configurations. Thousands of gear-shafts, camshafts, and power transmission rods would have enabled calculations carried out in one part of the machine to be relayed to other parts. Taken together, a completed Analytical Engine would have been mind-boggling in size, concept, and in the functions it carried out.

Above all, *its entire operation would have been controlled by precisely the same kind of cards that Jacquard used for his revolutionary loom.*

Reliable proof exists of the exact day when Babbage hit upon the idea of making the principle of Jacquard's loom so fundamental to the operation of the Analytical Engine.

The struggle Babbage had to come to terms with the magnitude of his vision of the Analytical Engine can be surmised from the fact that it took him two years from his initial vision of the Analytical Engine to decide precisely what control system he should use. In Volume II of the notebooks that Babbage himself described as his 'Scribbling Books'—his extensive hand-written journals preserved at the Science Museum in London—there is an entry for 30 June 1836 which contains the brief but momentous comment:

Suggested Jacard's loom as a substitute for the drums.

This was the first known time he mis-spelt Jacquard's name. As we have seen, it wasn't the last.

What exactly does the entry mean?

It reveals that prior to seeing that the Jacquard cards were ideal for operating his Analytical Engine, Babbage had toyed with the idea of programming his new machine by using a revolving drum featuring little raised studs as a mechanical means of inputting data and operating the machine. This type of drum was, of course, the basis for the control system of Jacques de Vaucanson's loom.

The notebook entry of 30 June 1836 marks the decisive moment when Babbage abandoned the idea of using the drum in favour of the Jacquard punched-card system. It was as though the evolution of the best method for controlling an automatic silk-weaving loom had been re-enacted in Babbage's mind. And, just as had happened in the silk-weaving industry itself, the Jacquard cards had won a resounding success over the de Vaucanson revolving studded drum.

Certainly, it made very good sense for Babbage to plump for the Jacquard cards. For one thing, producing punched cards is much easier and far cheaper than manufacturing studded drums. Yet, more importantly, the Jacquard control system offered the possibility of a potentially *limitless* programming system, whereas a revolving drum will, by definition, start to repeat itself before long. Yet Babbage did plan to use studded drums to control some internal processes of the Analytical Engine.

Babbage was a prolific writer on industry and machinery. It is possible that Babbage had known about the Jacquard loom since his university days, but in any event that it is likely that by 1836 he would have been familiar with the machine and with how it operated. In London's Spitalfields district a silk-weaving industry had sprung up that offered far from merely token competition to the silk industry in Lyons itself.

Writing in his 1864 autobiography, Babbage makes explicit the enormous influence and importance of the Jacquard loom:

> It is known as a fact that the Jacquard loom is capable of weaving any design which the imagination of man may conceive. It is also the constant practice for skilled artists to be employed by manufacturers in designing patterns. These patterns are then sent to a peculiar artist, who, by means of a certain machine, punches holes in a set of pasteboard cards in such a manner that when the cards are placed in a Jacquard loom, it will then weave upon its produce the exact pattern designed by the artist.

Now the manufacturer may use, for the warp and weft of his work, threads which are all of the same colour; let us suppose them to be unbleached or white threads. In this case the cloth will be woven all of one colour; but there will be a damask pattern upon it such as the artist designed.

But the manufacturer might use the same cards, and put into the warp threads of any other colour. Every thread might even be of a different colour, or of a different shade of colour; but in all these case the *form* of the pattern will be exactly the same—the colours only will differ.

The day when Babbage decided to make use of the Jacquard cards in his design for his Analytical Engine is one of the most momentous in our story. It is, literally, the day when the bridge between the weaving industry and the embryonic information technology industry was created. Babbage's decision was the most explicit confirmation, by the man who is today regarded as the father of the computer, of the argument at the very heart of this book: *that in essence a computer is merely a special kind of Jacquard loom.*

Babbage recognized that Jacquard's *automatic* use of the punched card as the means to control the raising and lowering of the warp threads on the loom for weaving brocade was a development of massive importance. After all, Babbage was basically trying to build a computer program a century before the word 'program' acquired this meaning, and at a time when 'computers' were—as we have seen—people, not machines. He found what he was looking for in Jacquard's cards. The moment he did, the global information revolution that is such a major part of our lives today took its first substantial step towards incarnation.

Bruce Collier, in his superb study of Babbage's work, *The Little Engines that Could've*, makes the important comment:

The introduction of punched cards into the new engine was important not only as a more convenient form of control than the drums, or because programs could now be of unlimited extent, and could be stored and repeated without the danger of introducing errors in setting the machine by hand: it was important also because it served to crystallise Babbage's feelings that he had invented something really new, something much more than a sophisticated calculating machine.

Collier's point here is of crucial importance. Furthermore, *today our whole idea of how computers should be programmed and even what they should actually be can be traced directly back to the Jacquard loom and its punched cards*. Indeed, the Jacquard card can even be said to constitute the invention of the binary digit or 'bit'.

A bit in computing terms is the smallest and most fundamental element of computerized information. This basically means that a bit is a unit of information expressed as a choice between two equally probable alternatives. Computers can be built because these alternatives can be boiled down to '0' or '1', alternatives that can in turn conveniently be represented electronically within the actual physical structure of a computer's circuitry by a tiny electronic switch that is either 'off' (for 0) or 'on' (for 1).

At the most fundamental level, this is how computers work. And the whole idea was essentially Jacquard's.

Writing in his autobiography, Babbage explains how the Analytical Engine would operate. He states that the machine would consist of two parts. These are, firstly, the *store* containing 'all the variables to be operated upon', and, secondly, the *mill* 'into which the quantities[1] about to be operated upon are always brought'.

Babbage's use of the terms 'store' and 'mill' are far-sighted anticipations of the modern computer features of computer memory and computer processor, respectively. In choosing the terms he did, Babbage was also alluding to the cloth industry of Totnes, the town where he spent much of his childhood and to which he often returned as an adult. As the distinguished com-

---

[1] In his autobiography Babbage writes 'qualities' not 'quantities' but this must be a misprint.

puter historian Martin Campbell-Kelly observes in his introduction to the 1994 reprint of Babbage's autobiography:

> This terminology was an elegant metaphor from the textile
> industry, where yarns were brought from the store to the mill
> where they were woven into fabric, which was then sent back
> to the store. In the Analytical Engine, numbers would be
> brought to the store from the arithmetic mill for processing,
> and the results of the computation returned to the store.

In his 1864 autobiography, Babbage points out that every formula
the Analytical Engine may be required to compute consists of
certain algebraic operations to be performed upon given letters,
and of certain other operations depending on the numerical value
assigned to those letters. By 'letters' Babbage is referring to letters in algebraic formulae such as $2x = 1$; $2y^3 = 16$, etc. although
the machine was designed to handle far more complex formulae
than this. He adds:

> There are therefore two sets of cards, the first to direct the
> nature of the operations to be performed—these are called
> operation cards: the other to direct the particular variables on
> which those cards are required to operate—these latter are
> called variable cards.

Strictly speaking, Babbage was specifying *three* types of cards,
because in practice there were also cards that contained the
values to load into the machine. In other words, some cards (the
Operation Cards) were to be used to control the actual *operations*
of the machine, others (the Variable Cards) to specify *from where
in the store the number to be operated on was to be fetched*, and still
others (the Number Cards) were to specify *the actual numbers on
which the machine operates*. A modern computer program works in
essentially exactly the same way.

Babbage concludes by making the observation that the
Analytical Engine is 'a machine of the most general nature'. He
explains that whatever mathematical formula it is required to

calculate, the details of the formula must be communicated to it by two sets of cards, and that once the machine has been programmed by the cards in this way, the Engine will always operate according to that formula until a new program is fed into it.

Babbage does not himself use the words 'programming' or 'program'. These terms had not yet entered the language and

How the Jacquard loom inspired Charles Babbage. These cards, amazing predecessors of the punched cards used to program the computers of the twentieth-century high-tech revolution, were in essence Jacquard cards, but used to weave calculations, not images. They were designed by Babbage.

he is therefore obliged to resort to more obscure expressions. For example, he describes the Analytical Engine as being made 'special' for the mathematical formula in question. In precisely the same way, we could visualize a Jacquard loom that was programmed to weave a lily as being made 'special' for the task of lily-weaving.

Almost as far-sighted as Babbage's entire concept of the Analytical Engine was his perception of it as having, in effect, a mind of its own. In a fascinating anticipation of the tendency of our own age to invest computers with a will and personality of their own ('I can't seem to do this; the computer doesn't like it'), Babbage consistently writes about the Analytical Engine as if it were a separate, reasoning entity. For example, in the chapter in his autobiography about the Analytical Engine he observes at one point:

> Thus the Analytical Engine first computes and punches on cards its own tabular numbers. These are brought to it by its attendant when demanded. But the engine itself takes care that the *right* card is brought to it by verifying the *number* of that card by the number of the card which it demanded. The engine will always reject a wrong card by continually ringing a loud bell and stopping itself until supplied with the precise intellectual food it demands.

Babbage also borrowed from the Jacquard loom the plan of creating what he describes as a 'library' of cards that carry out different functions, with the Analytical Engine's operator being able to take cards from the library as required and input them into the machine in order to make it special for the task. The enormous advantage of the Jacquard loom was, of course, precisely that it was able to weave any picture or pattern for which a chain of cards had been made. Weavers would keep these chains of cards in a storeroom whose function was very much the same as that of the library—or we might even say software library—which Babbage was proposing to create.

Babbage planned the Number Cards to express numbers in size up to $10^{50}-1$: that is, 9 followed by forty-nine 9s. In principle each number would be represented by columns, with as many holes punched in each column as were necessary to represent the units, tens, hundreds, thousands, and so on.

Babbage, who always did his utmost to maximize the practical usefulness of the Engines, envisaged a variety of methods in order to simplify how the machine would handle large numbers. These methods involved the Analytical Engine incorporating a variety of card-counting operations, with special holes being used in some cards to indicate that a number exceeded a particular quantity, or that the number itself was negative. The machine was designed to 'read' the cards just as the Jacquard loom did. Metal rods in the Analytical Engine mechanism would press against the cards. A particular hole would only register if the rod could poke through it.

Another inspired aspect of Babbage's punched-card system was that it provided a *permanent* record of the machine's readout. Furthermore, if the operator was certain that the correct series of punched cards had been installed in the machine at the start of the calculation, the operator could be confident that the *entire calculation* would be done without error.

Babbage also developed his own methods for reducing the number of operations needed to perform a particular calculation. Some of these are so complex that even today they are not fully understood. In addition, he designed systems by which his Engines would be self-correcting in the event that anything went wrong with them. In particular, his plans included measures that would render the machine unable to act on instructions from, say, a card that had inadvertently slipped out of place. These techniques used a system of locking devices that immobilized certain wheels during a calculating cycle, so that they would not be at risk of accidental movement. Babbage also designed mechanisms to ensure that a wheel could only be moved by an input from a legitimate source. He stated specifically in his plans that when the machine was working properly, it would be impossible for an operator to take any action that would produce false results.

As Babbage applied and extended Jacquard's ideas, he developed a passion for finding out all he could about the life and work of the French inventor. The first and most important consequence of this was the letter, already quoted, which he wrote in December 1839 to his friend Jean Arago asking the Frenchman to obtain two woven portraits of Jacquard if they were commercially available, or at least one if they were not. In the same letter, as we have seen, he also begged Arago to send him any 'memoir' that might shed some light on Jacquard's life.

A few weeks later, on 24 January 1840, Babbage received a reply in French. Arago wrote:

> My dear friend and colleague
>
> I fear that the person from Lyons of whom I have made enquiries for information about the Jacquard portrait must be out of town. I haven't had any answer to my queries ... Please be assured that I will completely fulfil, *con amore*, the commission with which you have charged me. I do not want you to have the slightest reason to doubt the high esteem in which I hold your talents and your character, nor the importance I attach to our friendship.
>
> Your devoted friend Jean Arago.

Arago was as good as his word. He stuck to the task, and by the spring appears to have been successful in obtaining at least one of the woven portraits that Babbage longed to own.

Driven on by curiosity and admiration, Babbage made a personal pilgrimage to Lyons later that year to see Jacquard's

Babbage's own invoice for the portrait he so much desired. This clearly shows the sum he paid— 200 francs.

loom in action. His visit to Lyons is yet further evidence of how intimately he felt his own work to be related to Jacquard's.

The story of Babbage's visit to Lyons contains some intriguing surprises. Buried in the Babbage papers at the British Museum there is an invoice, dated 8 September 1840, issued by the French Society for the Manufacture of Fabrics for the Furnishing and Decoration of Churches. This relates to the purchase of a 'tableau' (that is, the woven portrait) of Jacquard produced by the Lyons firm of Didier Petit & Co. The invoice is made out to 'Monsieur Babbage'. It seems quite clear that Babbage kept the invoice as a record of having purchased the woven portrait and of how much it cost him.

The firm of Didier Petit & Co no longer exists. The official guidebook for Lyons of 1835 reveals the company to have been a manufacturer of rich fabrics used in furnishing and church decoration. The address is given as 34 Quai de Retz. This is now known as Quai Jean Moulin (after the heroic World War II Resistance leader) and is located on the east side of La Presqu'île, just south of Croix Rousse. The invoice is for 200 francs. The daily average wage of an artisan in 1840 was about four francs. Comparing wage rates today with 1840, we can conclude that Babbage paid about £2500 (about $4000) at modern prices for his woven portrait of Jacquard—the clearest indication of how badly he wanted it.

It is natural to assume that the invoice in Babbage's papers relates to the woven portrait of Jacquard that Babbage obtained through Arago, and which he put on show at his soirées. But in fact this was not the case. Instead, the invoice turns out to be for a *second* portrait of Jacquard that Babbage obtained under rather more exotic circumstances.

In June or July 1840, Babbage had been invited by the Italian mathematician Giovanni Plana to attend a meeting of Italian scientists scheduled to take place in September in Turin, then the capital of the Kingdom of Sardinia. Babbage was invited to a similar meeting the previous year but had declined, pleading that he was too busy with his work on the Analytical Engine. This time he accepted. Very likely he did so because of the extraordinary insight into the importance of the Analytical Engine shown by Plana in his letter of invitation.

In his autobiography, Babbage recalled:

In 1840 I received from my friend M. Plana a letter pressing me strongly to visit Turin at the then approaching meeting of Italian philosophers. M. Plana stated that he had enquired anxiously of many of my countrymen about the power and mechanism of the Analytical Engine. He remarked that from all the information he could collect the case seemed to stand thus: 'Hitherto the *legislative* department of our analysis has been all-powerful—the *executive* all feeble. Your engine seems to give us the same control over the executive which we have hitherto only possessed over the legislative department.'

Considering the exceedingly limited information which could have reached my friend respecting the Analytical Engine, I was equally surprised and delighted at this exact prevision of its powers.

Plana's comment in effect amounted to a recognition that the Analytical Engine might be able to solve the long-standing problem of the lack of processing power to evaluate complex mathematical formulae. It was an extraordinarily far-sighted observation, and it is hardly surprising that Babbage was so thrilled at Plana's perceptiveness.

The German Romantic poet and philosopher Novalis once remarked: 'It is certain my conviction gains infinitely, the moment another soul will believe in it.' This could be a motto for all of Babbage's life; it explains much of his behaviour, especially during the long and often lonely years when he was labouring on his cogwheel computers. With no efficient working version of a Difference Engine or Analytical Engine to show the world, he was obliged to seek what seemed the next-best thing; the society of those who appeared to understand what he was trying to do. The fact that he was prepared to travel all the way to Italy—a far from easy journey in 1840, even for a man of Babbage's financial resources and energy—suggests how cut off from empathy and support at home he perceived himself to be.

Very possibly he was also influenced in his decision to make the journey by the fact that the journey to Turin offered an ideal opportunity to visit Lyons on the way, and find out more about Joseph-Marie Jacquard. The Lyons silk industry had sprung up there partly because of the city's proximity to Italy, and now Babbage was exploiting that very fact to combine his excursion to Turin with a visit to Lyons.

Babbage left England for Paris in the middle of August 1840. In the capital, he collected letters of introduction from Arago and other friends to people in Lyons.

Towards the end of August he arrived in Jacquard's birth-place. As he relates in *Passages*:

> On my road to Turin I had passed a few days at Lyons, in order to examine the silk manufacture. I was especially anxious to see the loom in which that admirable specimen of fine art, the portrait of Jacquard, was woven. I passed many hours in watching its progress:

What Babbage says here is ambiguous. Does he mean that he spent the 'many hours' just watching a Jacquard loom operating, or that he actually watched the loom weaving a 24 000-card Jacquard portrait? There is the tantalizing implication that the latter is the case.

If he had watched the portrait being woven, it would indeed have been an undertaking requiring 'many hours'. Assuming that the weaver was working at the usual Jacquard loom speed of about forty-eight picks per minute (say, 2800 per hour), the entire weaving process for the 24 000 picks would have taken more than eight hours per portrait, excluding breaks at the local *bouchon*. When Babbage was immersed in an intellectual pursuit, the intensity of his concentration was beyond compare. Whether he did in fact watch a Jacquard portrait being woven, is it not at least perfectly possible to *imagine* him observing the creation of the woven portrait of his hero from the very first pick of the shuttle to the very last?

And what about the invoice? The truth is that this relates not to the Jacquard portrait Arago procured for Babbage, but rather to a *second* woven portrait that Babbage obtained while in Lyons. This portrait, which he quite possibly witnessed being made, he did not keep for himself. Instead, later in his travels he made a present of it, in Turin, to the Queen of Piedmont and Sardinia, whose brother the Grand Duke of Tuscany, Leopold II, had extended friendship and hospitality to Babbage twelve years earlier, when Babbage visited Italy on the European tour he made to console himself after his wife Georgiana's death.

·8·
A question of faith and funding

In public Babbage kept up a bold front but his private letters tell a different story. He had little hope of an Analytical Engine being built in his lifetime. The government of the Duke of Wellington and the reform government had come and gone, and with them all hope of securing the necessary support for such a project in nineteenth century Britain. With no expectation of public support or backing, the courage and determination with which he pursued the work in the face of enormous difficulties is very impressive.

Anthony Hyman,
Charles Babbage, Pioneer of the Computer, 1982

When we think of Babbage leaving for Turin to discuss the Analytical Engine with Giovanni Plana and other leading Italian scientists in the summer of 1840, and meeting royalty, it is easy to romanticize the situation and assume Babbage was at the pinnacle of his achievements.

But this was not the case at all. The truth is that when Babbage left for Italy, his career at home was in shreds. Yes, among some of Europe's greatest minds, Babbage could look forward to receiving the respect and understanding he longed for. Back in Britain, though, negotiations with the Government over the provision of funding to continue his work had foundered.

This was to a large extent because Babbage was finding it increasingly difficult to get taken seriously in his native land. He was a genius, but diplomacy was not his strong point. Back in 1834 he had made the fatal, and really rather stupid, mistake of telling the British Government he had abandoned work on the Difference Engine because he had invented another machine which 'superseded' it. The other machine was, of course, the Analytical Engine.

Why did Babbage inform the Government he had abandoned work on the very machine the Government had supported with such lavish financial grants? The confession did not even turn out to be strictly true: he never entirely abandoned his labours on the Difference Engine and was still conducting useful work on it in the 1850s. But Babbage was obsessively honest and throughout his life motivated by a sense of justice so pronounced that it often placed his own interests in jeopardy. He felt it was, in effect, only fair to mention the new direction his work was taking. Another likely factor was that he was proud of his new idea and keen to tell people about it.

One sentence from his letter to the Government about his intentions reveals the strength of his sense of justice. Babbage explained that he was revealing his plans in detail in order that 'you may have fairly before you all the facts of the case'. His mistake also stemmed from his almost incredible naïveté in political matters. He lacked the politician's essential skill of having a canny understanding of the likely effect of a particular action or statement on an audience.

In any case, financial backers want good news, not honest but unpleasant revelations. Under the circumstances the Government could hardly be blamed for putting its own sense of justice to work and concluding that Babbage was behaving with utter irresponsibility. The fact was that the British Government had, as we have seen, spent £17470 on the Difference Engine. Many people in the Government responded with indignation, suspicion, and even fury to the news that work had stopped on the Difference Engine. There had always been those in the corridors of power, and in the scientific world generally, who found it convenient and easy to regard Babbage as a troublesome fraud, extorting money from the Exchequer. Some even put it about that he was dishonestly using the money to prop up his own lifestyle.

When Babbage announced that he had abandoned the Difference Engine, his detractors and enemies set to work on his reputation with knives and cudgels. The Government did not at this stage completely rule out offering Babbage new funding to work on the Analytical Engine, but neither did they indicate that this was—so to speak—on the cards. Babbage remained in a limbo, having no idea whether additional financial help might be forthcoming. The Analytical Engine had entered development hell.

Babbage returned to Britain in September 1840. Temporarily exhilarated by his visit to Turin, where he had been received as an eminent international inventor and man of science, he had to confront the depressing truth that it was increasingly unlikely he would ever be able to afford to build the Analytical Engine. He knew that his own financial resources, considerable as they were, would be nowhere near enough for the task. Yet somehow he continued to apply himself with energy and dedication to his great object.

He was sustained by two hopes. Firstly, he thought it reasonably likely that one of the Italian scientists whom he had recently visited might write a lengthy and detailed paper on the new project. Babbage hoped that such a paper would affirm the importance of the invention and give him leverage with the British Government. He had a curious habit of trying to win influence by these rather indirect means instead of honing his diplomatic skills and adopting a more direct, and possibly more successful, approach. His host in Turin, Giovanni Plana, had indicated that he himself was not in sufficiently good health to undertake the job, but Luigi Federico Menabrea, a talented young mathematician whom Plana had introduced to Babbage in Turin, appeared interested in carrying it out. Babbage remained in touch with Menabrea and supplied him with comprehensive information about the Analytical Engine.

Babbage's second cause for optimism, however tenuous, was that the Government might, after all, have a spontaneous change of heart and make new funding available to him.

Babbage went on working, as he always did, while continuing to press for a decision on whether the Government was prepared to give him more money. He had been hoping for more money since 1834, when he first had the idea for building an Analytical Engine. Now it was six years later, but in fact another two years were to elapse before he finally received his answer.

It arrived by post in the first few days of November 1842.

Sir Robert Peel, founder of the Conservative (Tory) party, whose members included landed gentry, wealthy industrialists, and the upper middle-class, had been elected Prime Minister the previous year. He could be sympathetic to men of science, but not when they made representations to the Government for money. Peel's view—one that has, after all, been shared by many governments since—was that if an invention was any good it would attract capital from the free market, and if it wasn't and didn't, it hardly deserved to be funded by the public purse. Peel loathed the idea that the Government should be expected to be the banker of last resort for desperate inventors whose ideas had failed to attract capital from any other source.

On 31 August 1842, Peel had written a confidential and scathing letter about Babbage to a friend, the geologist William Buckland. The letter makes all too clear what Peel thought of Babbage:

What shall we do to get rid of Mr Babbage and his calculating Machine? I am perfectly convinced that every thousand pounds we should spend upon it hereafter would be throwing good money after bad. It has cost £17 000, I believe—and I am told would cost £14 or £15 000 more to complete it.

Surely if completed it would be worthless so far as science is concerned?

What do men really competent to judge say of it *in private?*

It will in my opinion be a very costly toy to complete and keep in repair. If it would now calculate the amount and the quantum of benefit to be derived to Science it would render the only service I ever expect from it.

I fear a reference to the Royal Society, and yet I should like to have some authority for treating this Calculating Machine as I should like to treat the Caledonian Canal—and would have treated it but that I was told it would cost £40 000 to unravel the web we have spent so many hundred thousand pounds in weaving.

I will consider any opinion you may give me as to the course which should be pursued strictly confidential.

Pray read the enclosed papers.

Most truly yours

Robert Peel

Viewed from a pure cost perspective, Peel's caustic attitude to Babbage was, perhaps, entirely reasonable. But Prime Ministers are elected to be leaders and visionaries, not accountants. Babbage was planning a new kind of revolution—an information technology revolution, and all Peel could think of was his Exchequer.

Yet Peel was much too canny an operator to make public the strength of his contempt for Babbage. He was only too aware, after all, that Babbage was one of the leading intellectuals in England, with influential friends in science, business, and politics, many of whom were leading Conservatives. In particular, Peel was aware that Babbage was good friends with the Duke of Wellington and Prince Albert.

The Duke, by now an especially close friend of Babbage, had been a prime mover behind Babbage securing funding for the Difference Engine back in the 1820s. Indeed, it is ironic that Babbage received *all* his official funding for his visionary cog-wheel calculator from supposedly backward-looking aristocrats (who can perhaps be compared, very loosely, to the noblemen of pre-Revolutionary France) but didn't receive a penny from the upper middle-class politicians who attained power after the Reform Act of 1832. The Reform Act broadened the electoral base and gave the towns and cities much more extensive parliamentary representation.

As cunning as the foxes he loved to hunt in what little spare time he had, Peel decided to place the burden of the Babbage funding decision on someone else's shoulders. In effect, he passed the buck. He commissioned a report on the Difference Engine from the Astronomer Royal, George Biddell Airy.

Airy had been appointed Astronomer Royal in 1835, at the early age of thirty-four. By then he had already made many valuable contributions to science, especially in the field of optics. He was to remain Astronomer Royal until 1881, yet he was never regarded as having distinguished the appointment.

Later in his career, in 1846, he was vilified by the public for failing to act on the findings of a young English astronomer, John Couch Adams. Adams concluded in 1845, after laborious calculations, that there were certain irregularities in the orbit of Uranus suggesting the possible existence of a new planet. Instead of instigating a major telescopic search that would almost certainly have resulted in the discovery of the new planet, Neptune, Airy chose not to act on Adams's information. As a result the planet was instead discovered in 1846 by a German astronomer, Johann Gottfried Galle.

There was every reason on Earth why Airy should have been exactly the right person to understand just how much potential importance there was in Babbage's ideas. In the course of his astronomical work, Airy had to carry out a large number of complex, repetitive arithmetical calculations. But the truth was that Airy was less a scientist than a bureaucrat. He was also a pedantic, narrow-minded man with little or no real vision of what science could achieve for humankind. To make matters worse, he detested Babbage, perhaps because in his heart of hearts Airy knew that he himself did not entirely understand what Babbage was trying to achieve. Airy was no fool, but Babbage was operating at the very limits of scientific and mathematical possibility, and only those with the most fearless and visionary intellects were likely to grasp exactly what he was seeking to do. In any event, Airy took every opportunity to attack Babbage's plans for cogwheel calculating machines.

There is little doubt Peel had a pretty good idea, in advance, of what Airy thought of Babbage. Peel rarely missed a trick in the political sphere, and his network of contacts made even Babbage's look slim. Asking Airy to report on Babbage's work was like asking someone who was obsessed, say, with the transportation potential of canals to prepare a report on the future of the railway, which competed heavily with the canals in the mid-nineteenth century.

When Airy delivered his report, Peel was certainly not disappointed. Airy's crass judgement on the Difference Engine was that it was 'useless'.

And so, probably on Thursday 3 November 1842 (at that time letters were often delivered in London on the same day they were posted), Babbage received a letter from the Chancellor of the Exchequer, Henry Goulburn.

3 November 1842
Downing Street

My Dear Sir

The Solicitor General has informed me that you are most anxious to have an early and decided answer as to the determination of the Government with respect to the completion of your calculating Engine. I accordingly took the earliest opportunity of communicating with Sir Robert Peel on the subject. We both regret the necessity of abandoning the completion of a machine on which so much scientific ingenuity and labour has been bestowed. But on the other hand the expense which would be necessary in order to render it either satisfactory to yourself or generally useful appears on the lowest calculation so far to exceed what we should be justified in incurring that we consider ourselves as having no other alternative. We trust that by withdrawing all claim on the part of the Government to the machine as at present constructed and by placing it at your entire disposal we may to a degree assist your future exertions in the cause of science.

I have the honour to be
Dear Sir
Yours ever most faithfully
Henry Goulburn

Sir Robert Peel begs me to add that as I have undertaken to express to you our position now on this matter he trusts you will excuse his not separately replying to the letter which you addressed to him on the subject a short time since.

Having now met Babbage, we know him well enough to be pretty certain that he *wouldn't* excuse Peel this, and he didn't. Babbage's anger at the contents of the letter was compounded at being fobbed off with a subordinate. On 6 November he demanded an interview with the Prime Minister. On Friday, 11 November, at eleven o'clock in the morning, he was granted one.

The interview was, to put it mildly, an unmitigated disaster for Babbage. It is possible to reconstruct it almost on a minute-by-minute basis from a detailed account Babbage wrote of the interview. The word 'wrote' is in fact not really adequate to describe how it came to be composed. What happened was that immediately after the interview, in a hot fury of anger and

Sir Robert Peel, pugnacious, politically brilliant, was no judge of Babbage's ideas.

disappointment, Babbage rushed back into his house, dashed into his study and—as if aware this was the only way he could obtain any relief—gouged onto paper a blow-by-blow account of what he must even at the time have realized was a meeting that pretty well killed his twenty-year vision of cogwheel computing stone dead.

The document containing the account is lodged in the British Library in London. Holding the document in one's hands is an experience at once deeply moving and profoundly troubling. After all, what Babbage wrote was no less than an explanation of how the information technology revolution, that really might have happened in Victorian Britain, was in fact suffocated at birth. Babbage's vivid account of the meeting includes much of the verbatim dialogue that passed between the two men. His pain and upset are even apparent in the appearance of the writing itself, which is erratic, contains many crossings out, and is not always easy to decipher. Uncharacteristically for him, in his haste and anger his description of the meeting leaves out much of the punctuation and even some of the words.

The timing of the meeting was, from any perspective, extremely unfortunate. The year 1842 had been an exceptionally tough one for Peel. Shortly before the day when he met Babbage, he had written to his wife, Julia, that he was 'fagged to death' with the cares of office. Hunger and rioting were widespread. Peel was in no mood to meet Babbage at all, let alone for a stressful confrontation. Even so, Babbage would surely have done better if he had handled the meeting in a radically different manner. He should have been placatory, and done his utmost to explain to Peel about his work in layman's language. As it was, and this is clear even from Babbage's own notes, he conducted the meeting in a defensive, sullen, bad-tempered, and self-pitying manner that only served to irritate and alienate Peel. As Babbage explains in his account of the interview:

> I informed Sir RP that many circumstances had at last forced upon me the conviction, which I had long resisted, that there existed amongst men of science great jealousy of me. I said that I had been reluctantly forced to this conclusion of which I now had ample evidence, which however I should not state unless he asked me. In reply to some observation of Sir RP in a subsequent part of the conversation I mentioned one circumstance that within a few days the Secretary of one of the foreign embassies in London has incidentally remarked to me that he had long observed a great jealousy of me in certain classes of English Society.

(*right*) Hot off the press. This first page of the memorandum which Babbage scrawled in a fit of fury after his disastrous interview with Robert Peel shows the painful deterioration in Babbage's usually elegant handwriting as a result of his anger.

Recollections of an interview with
Sir R. Peel on friday Nov 11 — 1842   at 11 a.m.

I first ~~xxxxxxxxxx~~ asked Sir Peel if he
had seen Sir Wm Follett within the last two days
assigning as my reason that if he had it might
perhaps save him ~~xxx~~ time as in that case Sir Wm
Follett would have anticipated from me some part of what I
had to communicate.

I then informed Sir R.P. that many circumstances
had at last forced upon me the conviction which I
had long resisted that there existed amongst men of
science great jealousy of me ~~xxxxxxxxxxxxxxx~~ ~~xxxxxxxxxxx~~. I said that I had been reluctantly
forced to this conclusion of which I now had ample
evidence which however I should not state unless
he asked me — In reply to some obs.n of Sir R.P.
in a subsequent part of the conversation I mentioned
one circumstance that within a few day the
secretary of one of the foreign ~~xxxxx~~ embassies
in London had incidentally remarked to me
that he had long observed a great jealousy of science
in certain classes of English society.

I then said that as Sir P.d must of course obtain
his views both of the Difference and the Analytical
engine from others I thought it right with not
wishing to allude to any individuals ~~xxx~~ not ~~xxxx~~
wishing to know the name of any of his scientific
advisers yet for his P.s sake as well as my own
to ~~xxx~~ mention this conviction. ————

I then turned to the next subject the importance of the Anal Engine   I stated my own
opinion that in the future scientific history of
the present day it would probably form a marked
epoch. and that much depended upon the result
of this interview .   I added that the Difference

Babbage went on to explain why he was mentioning all this. He told Peel of his fears that some of those who had advised the Government over the worth of the Engines might have based their decision on personal malice rather than on an objective assessment. Peel evidently made no direct reply to this. Then, finally, Babbage got down to what really mattered:

> I turned to the next subject, the importance of the Analytical Engine. *I stated my own opinion that in the future scientific history of the present day it would probably form a marked epoch and that much depended upon the result of this interview.* I added that the Difference Engine was only capable [of] applications to one limited part of science (although that part was certainly of great importance and capable of more immediate practical applications than any other) but the Analytical Engine embraced the whole [of] science.
>
> I stated that it was in fact already invented and that it exceeded any anticipations I had ever entertained respecting the powers of applying machinery to science.

The italics in the above passage are mine. The brilliance of Babbage's anticipation of how posterity would see the Analytical Engine is deeply moving and somewhat uncanny, almost as if he were at that moment a time-traveller with a personal experience of the future. Yet as far as his dealings with the pragmatist Peel were concerned, Babbage ought to have given him a clear indication of the practical benefits of his machines, accompanied by a realistic plan of action and a date when the Government could expect that something of definite practical usefulness would be completed. But Babbage, in his bitterness and haste to justify himself, was not thinking clearly at all. He tried quoting to Peel the comment Plana had made that the invention of the Analytical Engine would provide 'the same control over the executive [department of analysis] as we have hitherto had over the legislative'. Peel very likely did not have the faintest idea what Babbage was talking about.

After another bad-tempered and unpleasant foray, initiated by Babbage, about the different pensions and grants given to scientists by the Government, Peel finally decided to interrupt the endless stream of complaints and grievances with a hard fact:

'Mr Babbage, by your own admission you have rendered the Difference Engine useless by inventing a better machine.'

Babbage, recalled to reality, glared at Peel. 'But if I finish the Difference Engine it will do even more than I promised. It is true that it has been superseded by better machinery, but it is very far from being "useless". The general fact of machinery being superseded in several of our great branches of manufacture after a few years is perfectly well known.'

Only briefly diverted from his course, Babbage again went on to complain of all the vexation and loss of reputation he considered that he had suffered from those members of the public who believed him to have profited personally from the money the Government had granted towards the development of the Difference Engine. 'This belief is so prevalent that several of my intimate friends have asked if it were not true,' Babbage said. 'I have even met with it at the hustings at Finsbury.'

Peel was on home territory now. 'You are too sensitive to such attacks, Mr Babbage,' he replied. 'Men of sense never care for them.'

Fixing the Prime Minister with another hard stare, Babbage returned, 'Sir Robert, in your own experience of public life you must have frequently observed that the best heads and highest minds are often the most susceptible of annoyance from the injustice or the ingratitude of the public.'

The fact of two of the most eminent men of the nineteenth century holding such an unproductive and ill-tempered conversation leaves us in no doubt that neither of them was at his best that day. Peel was exhausted, and irritated with Babbage. Babbage felt hurt, betrayed, and angry that the Prime Minister could be so reluctant to support the continued development of machines whose worth seemed to Babbage self-evident. Babbage asked Peel to contemplate the dream; Peel saw only a highly unsatisfactory reality.

'I consider myself to have been treated with great injustice by the Government,' was Babbage's parting comment. 'But as you are of a different opinion, I cannot help myself.'

Babbage got up from his chair, wished his Prime Minister good morning, and abruptly left the room.

Hurt and humiliated by the meeting, Babbage was hardly likely to want to write about it or generally make his hurt and humiliation public. But very likely he could not help speaking about it to his closest friends. One of these, the great novelist Charles Dickens, appears to have drawn on Babbage's disappointment at the hands of the British Government in his 1855 novel *Little Dorrit*, which is a scathing indictment on the innate cruelty of official institutions.

From the years 1839 to 1851, Babbage and Dickens lived only a few hundred yards from each other; Dickens in his large house on Devonshire Terrace in Marylebone Road, and Babbage in Dorset Street. They probably met in 1838 through their mutual friend the actor William Macready. Babbage and Dickens moved in similar circles, and were often guests at each other's dinner parties.

Dickens was most certainly not of a scientific disposition or frame of mind, and had little or no technical knowledge of Babbage's work. But he had no problem understanding the benefit to mankind and freedom from mental drudgery that a calculation machine would bring. Writing from Broadstairs, Kent, to his brother Henry Austin on 20 December 1851 about the rocketing costs of the modifications to his new house in London's Tavistock Place, Dickens was ruefully and ironically to comment that the bill submitted by the builder was 'too long to be added up, until Babbage's Calculating Machine shall be improved and finished ... there is not paper enough ready-made, to carry it over and bring it forward again'.

A crucial theme in *Little Dorrit* is how the cold and indifferent workings of the law and government bring human misery. The tenth chapter of the first part of the book, entitled with transparent irony 'Containing the Whole Science of Government', focuses on a Government department dedicated to never getting anything done. Dickens calls it the 'Circumlocution Office':

> The Circumlocution Office was (as everybody knows without being told) the most important Department under government. No public business of any kind could possibly be done without the acquiescence of the Circumlocution Office ... Whatever was required to be done, the Circumlocution Office was beforehand with all the public departments in the art of perceiving—HOW NOT TO DO IT ... Through this delicate perception, through the tact with which it invariably seized it, and through the genius with which it always acted on it, the Circumlocution Office had risen to overtop all the public departments; and the public condition had risen to be—what it was.

One of the most put-upon victims of the Circumlocution Office is an inventor called Daniel Doyce. Dickens describes him as a 'a quiet, plain, steady man', who 'seemed a little depressed, but neither ashamed nor repentant'.

We are told that a dozen years earlier, Doyce has perfected 'an invention (involving a very curious secret process) of great importance to his country and his fellow creatures'. But instead of winning praise from officialdom for what he has done, from the moment Doyce approaches the Government for help with funding, he 'ceases to be an innocent citizen, and becomes a culprit. He is treated, from that instant, as a man who has done some infernal action.'

Dickens's imagination got by perfectly well, most of the time, without needing to use real people as the basis for all the characters he created. There were about 2000 of these, according to Peter Ackroyd, one of his most popular modern biographers.

The similarities between Doyce and Babbage, though, are too striking to be ignored. The description of Doyce's appearance is a good fit to Babbage to begin with ('... a practical looking man, whose hair had turned grey, and in whose face and forehead there were deep lines of cogitation') and even the timing of when Doyce 'perfected' his invention ('a dozen years ago') which after all is an entirely free choice on Dickens's part, seems to have been chosen very deliberately to allude to Babbage's work. As for the account of the invention itself, its 'great importance to his country and his fellow creatures' also seems to point directly to Babbage, as does the ironic account of Doyce's plight voiced by another character, Mr Meagles:

> '... he has been ingenious, and he has been trying to turn his ingenuity to his country's service. That makes him a public offender, sir.'

And what Doyce says about how inventors such as he are treated at home compared with abroad could easily have been words taken down pretty well verbatim from some lament Babbage might, in a self-pitying mood, have made at one of Dickens's numerous dinner parties at Devonshire Terrace, over the turtle soup, the turbot, and the roast lamb.

> 'Yes. No doubt I am disappointed. Hurt? Yes. No doubt I am hurt. That's only natural. But what I mean, when I say that people who put themselves in the same position, are mostly used in the same way—'
>
> 'In England,' said Mr Meagles.
>
> 'Oh! of course I mean in England. When they take their inventions into foreign countries, that's quite different. And that's the reason why so many go there.'

For Babbage, the conclusion to his meeting with Peel certainly meant the end of any realistic hopes he might have had for completing the Difference Engine in England, let alone for building the massively more ambitious Analytical Engine. It is no exaggeration to say that Airy's damning report on the Difference Engine, and Peel's decision to accept his recommendation, put paid to any prospect of a cogwheel-based information technology revolution taking place in Britain in the 1840s.

Yet, there was no inherent reason why Babbage's machines should not have formed the technology for such a revolution in the 1840s or 1850s. If ever there was an age which deployed remarkable energy in planning and advancing the practical implementation of innovative technology, it was the nineteenth century.

A confidence in the future, a conviction that life was getting better, a belief that man not only has the potential for happiness and self-fulfilment but has a duty actively to pursue these destinies; all these lay at the heart of so much nineteenth-century thinking. When a new and useful technology became practical and feasible it was invariably pressed into service without delay. This spirit of optimism governed thought and action. Above all, the age believed in the power of machinery to achieve almost anything. As Thomas Carlyle—who invented the expression 'captains of industry' and who once remarked that man was above all 'a tool-making animal'—wrote in his 1882 retrospective essay *Signs of the Times*:

> Were we required to characterise this age of ours by any single epithet, we should be tempted to call it, not an Heroical, Devotional, Philosophical, or Moral Age, but, above all others, the Mechanical Age. It is the Age of Machinery, in every outward and inward sense of that word; the age which, with its whole undivided might, forwards, teaches and practises the great art of adapting means to ends.

Babbage's work ultimately furnished a spectacularly original way whereby precise mechanical means might be adapted to the end of making machines that generated an infinitely useful commodity: accurate, reliable, rapid calculations. As Babbage's friend and would-be biographer Henry Wilmot Buxton wrote in the 1870s of Babbage's work:

> The marvellous pulp and fibre of a brain had been substituted
> by brass and iron; he had taught wheelwork to think.

But beautifully expressed as this statement is, it is in truth more a statement of the success to which Babbage aspired than a reflection of what he actually managed to achieve when he was alive. He understood that the purpose of machines is to make up for the inherent limitations of human abilities. In his case, what he was most concerned about, as we have seen, was the alarming tendency for people to make mistakes when they undertook calculations due to the sheer mental drudgery these calculations required.

On the face of it Babbage should have had everything going for him. He was rich, he had a tremendous breadth of contacts and devoted admirers who included some of the most eminent people in the land. He had the ear of the Government, at least initially. Compare his position to that of Jacquard at the outset of *his* career, and it seems inconceivable that Jacquard would succeed and Babbage would not. Yet certain crucial factors, combined with the daunting technical challenges (which, in fairness to Babbage, were even greater than those faced by the Frenchman), resulted in Babbage's failure.

In particular, Babbage was severely handicapped by his stubbornness when it came to the standards of precision he wanted to attain. Modern research has demonstrated that the levels of precision Babbage aimed for were higher than he actually needed. With hindsight, this slowed him down significantly.

He was also hampered by his political and diplomatic naïveté. This, as we know, tended to lead to his inept handling of influential people who might have helped him.

Another serious problem was his unfortunate reluctance to work on equal terms with his brilliant engineer Clement. Babbage always treated him as a mere servant, albeit an important one.

Babbage was very far from being a snob—he was perfectly at ease socially with tradesmen and working people—but he had been born into the ruling-class of England and never quite forgot this. His attitude alienated Clement, who was not an easy person to get on with at the best of times. Finally, the two men's working relationship broke down completely, and Clement left Babbage's service.

The result of all this was that Jacquard enjoyed fame, glory, and fulfilment, while Babbage went to his grave a bitterly disappointed and disillusioned man.

If Babbage had successfully completed a working example of this first type of cogwheel computer, the Difference Engine, the resources and the money would surely have been made available for him to have at least a better-than-evens chance of completing the much more complex and ambitious Analytical Engine. Babbage himself never doubted that technology could achieve almost anything. If he had had a little more success in his endeavours and proved his machines financially viable, can we really doubt that the energetic, visionary, and money-obsessed age in which he lived would have made his dreams come true?

There would be little more to tell about Babbage's role in *Jacquard's Web* were it not for Ada Byron, Lord Byron's daughter. We have already briefly met her when looking at how Babbage explained the origins of his plans for the Analytical Engine to Ada, her mother Lady Byron, and their friend Mary Somerville.

Ada was one of the few people, apart from Babbage himself, who understood the full importance of Babbage's work in general and his invention of the Analytical Engine in particular. Aristocrat, amateur mathematician, housewife, occasional laudanum addict, and self-proclaimed genius, Ada was an ally of the most enthusiastic and supportive kind. And, most importantly for us, she was particularly fascinated by the relationship between the Analytical Engine and the Jacquard loom.

·9·

The lady who loved the Jacquard loom

Is thy face like thy mother's, my fair child!
Ada! sole daughter of my house and heart?
When last I saw thy young blue eyes they smiled,
And then we parted,—not as now we part,
But with a hope.—

             Awaking with a start,
The waters heave around me; and on high
The winds lift up their voices: I depart,
Whither I know not; but the hour's gone by
When Albion's lessening shores could grieve or glad mine eye.

            Lord Byron. *Childe Harold's Pilgrimage*,
                           Canto 3, 1816

If Ada Byron could somehow emerge from the tomb in the Nottinghamshire church in the English Midlands where she has lain since 1852 and get to know our world, what would she think of it?

119

She always adored technology. She would certainly be fascinated to see the extensive and indeed extraordinary technological advances that have taken place between her day and our own. Throughout her girlhood she dreamed of flying in a flying machine and even made tentative plans to build a steam-powered one. It would be a delight to take her on a flight in an airplane and let her experience her dream come true.

Most of all, though, in view of her fascination with Charles Babbage's work that lasted all her adult life, she would love to see how many computers there are in our world, how powerful they are, how *inexpensive* they are, and how small and compact. True, she might be disappointed to find how little they resemble Babbage's cogwheel machines in external appearance, but once Ada became acquainted with the range of functions and features they offer, she would surely be happy to overlook that. As, one feels, would Charles Babbage himself. After all, he only used cogwheels because they were the sole technology available in his own day that he could imagine to offer any possibility of allowing the creation of calculating machines. He certainly did not have any inherent, stubborn attachment to cogwheel technology.

Ada has become a famous figure in the history of computing due to her association with Babbage, who is with some justification regarded today as the father of computing. Ada was famous during her lifetime, too, but for a different reason.

Her celebrity in the nineteenth century stemmed from her father being the poet Lord Byron. For much of the century he was the most popular poet in Europe, as renowned for his omnivorous love life as he was for his passionate, verbally dazzling poems. He fathered many children in his sexual peregrinations around Britain and Continental Europe. Unlike his other children, however, Ada was born in wedlock, for Byron temporarily succumbed to marriage with Ada's mother.

The daughter of a great romantic poet, a Victorian housewife and mother, Ada was an unlikely collaborator with Charles

(*left*) Portrait of Ada Lovelace.

Babbage in his work on the Analytical Engine. Yet in fact, she turned out to be his most important ally. Today, her reputation in connection with the prehistory of the computer is growing in direct proportion to the ever-increasing importance of computers in our world. Here is Ada in action:

> In studying the action of the Analytical Engine, we find that the peculiar and independent nature of the considerations which belong to *operations*, as distinguished from *the objects operated upon* and from the *results* of the operations performed upon these objects, is very strikingly defined and separated.

The above passage is typical of the calibre of the work attributed to Ada. This passage could almost be an extract from a modern computer manual, *but written by a Victorian*. It is included in a Note that Ada appended to a translation she published in 1843 of a scientific paper, constituting a detailed discussion of the Analytical Engine, which had been published in French the year before. The paper Ada translated was written by the Italian scientist Luigi Federico Menabrea. He had finally come up with the goods. Menabrea's paper is a detailed account of the concept and purpose of the Analytical Engine. The paper was published in the October 1842 issue of *Bibliothèque Universelle*, one of Continental Europe's foremost scientific journals.

The passage just quoted provides a clear anticipation of the fundamental difference between computer *software* (that is, the set of instructions according to which a computer operates) and *data* (that is, the actual numbers on which the computer's actions are performed, and the results derived from these numbers). Ada is basically expressing the principle using nineteenth-century language.

And this, of course, is exactly the same idea Jacquard embodied in his loom. It is also the concept Babbage borrowed from Jacquard and incorporated into the Analytical Engine. Ada was particularly sensitive to the significance of the Jacquard loom in Babbage's plans. The best way to understand her contribution to

Babbage's work is to see it in terms of the relationship between Babbage's work and Jacquard's. Ada loved the Jacquard loom, and added it to her many scientific fascinations.

Menabrea's paper is heavily based on information Babbage supplied to him. We can be certain of this, because there are extensive similarities between much of Menabrea's material and such writing as Babbage himself did on the Analytical Engine.

Menabrea, for example, takes considerable pains to point out that the invention of the Analytical Engine was *not* the cause of the abandonment of the work on the Difference Engine. Babbage always makes the same point. Menabrea emphasizes that work on the Difference Engine had in fact already been set aside before Babbage started to develop the concept of the Analytical Engine. It is known that misconceptions about this matter greatly aggrieved Babbage, who took every step to encourage Menabrea to publicize what really happened. (As we have seen, Babbage did in fact subsequently again take up work on the Difference Engine in the 1850s, but Menabrea could not of course have known he would do that.)

Had Ada *only* translated Menabrea's paper her achievement would have merely been a linguistic one. Furthermore, one could reasonably have said that she was to a large extent translating a paper whose intellectual content was substantially Babbage's own. But in fact Ada's translation was merely the starting-point for her work, for her published translation is accompanied by seven additional Notes (she consistently capitalizes the word). These, denoted by the letters A to G, extend to more than 20 000 words: that is, about twice as long as the actual translation. They offer a penetrating insight into the Analytical Engine, with its revolutionary design and objectives and with its intimate conceptual connection with the Jacquard loom. As one reads the Notes, the information technology revolution seems just around the corner, instead of more than a century in the future.

Ada's accident of birth gave her a very special kind of life: it put her into the public spotlight from the moment she was born. In 1837, when Ada was twenty-one, Benjamin Disraeli, a novelist before he became a politician, could base the heroine of his novel *Venetia* on Ada and take it for granted that his readers would recognize the portrait.

Ada was raised by her mother, Lady Byron, born Annabella Milbanke. Annabella was a rich, spoilt, bossy, manipulative, but self-possessed and intelligent woman who had managed to tolerate married life with Lord Byron for just over a year. Byron's reasons for marrying Annabella are still rather a puzzle—indeed, many who admire his poetry would probably find it difficult to believe he was ever married at all. His most likely reasons for entering that state were that he hoped his wife's family's money would help to free him from his habitual indebtedness (it didn't) and because he was hoping Annabella would provide him with a son (she didn't).

Annabella and Byron were married on 2 January 1815, in her parents' manor house in the small town of Seaham, on the northeast of England. Things did not go well even during the honeymoon, during which Byron slept with a loaded pistol by his bedside. He also took care to inform his doting wife that he felt as though he were 'in hell'. For a couple of years prior to his marriage he had been having an on–off affair with his half-sister, whose name was Augusta. Ada herself, the fruit of a brief period of mutual affection between Byron and his wife near the start of the marriage, was christened Augusta Ada when she was born on 10 December 1815, but the 'Augusta' was soon dropped. By then her mother had started to suspect, apparently justifiably, the sexual nature of the liaison between Byron and his half-sister. This liaison was still continuing even after the marriage. Ada was only a few weeks old when Annabella, sick of Byron's unstable moods, debts, and infidelities, took the child and fled from her husband's London house in the middle of the night. Byron never saw his wife or his daughter again.

Byron was incapable of extending lasting love to anyone apart from himself, and even here he often faltered. If the real nature of his feelings towards his daughter seems an enigma, this is only to say that the real nature of all his feelings appears similarly enigmatic. It is not even clear whether the love he claimed to have for Ada was a genuine love or the kind of sentimental, remote, 'poetic' fondness he applied to many people he rarely saw, including the two illegitimate children he is definitely known to have fathered.

The mystery is hardly solved by the stanza, quoted at the start of this chapter, from the Third Canto of Byron's autobiographical poem *Childe Harold's Pilgrimage*. In this part of the poem, Byron sets down his feelings about Ada when he was sailing away from England after the failure of his marriage. There is evidently some fatherly affection here, but it also sounds as if father is not altogether unhappy to be leaving his daughter behind and rather relishes the prospect of being relieved from the more practical aspects of fatherhood and consequently being able freely to pursue his own agenda overseas. Lady Byron, however, fearing that if Ada ever left Britain, Byron might find the girl and kidnap her, only took Ada on a tour of Europe after Byron died, on 19 April 1824, of swamp fever at Missolonghi, Greece. He was thirty-six.

Ada inherited Annabella's love of learning and confidence in her own opinion. Fortunately, the excellent education her mother arranged for her to have at home tended to temper, though not remove, a tendency Ada had towards overconfidence in her intellectual abilities.

The usual educational opportunities open to girls in the early nineteenth century were limited, to put it mildly. Even middle-class and aristocratic girls were usually only taught such skills as were necessary for overseeing the management of the households they could one day expect to manage. Many professional educators, even female ones, actually regarded women's minds as inferior to men's at a fundamental biological level. Foreign literature, especially French plays and novels, were seen as particularly dangerous to women, possibly because it was feared it would open up emotional possibilities to women beyond the narrow path of marriage and life as a placid ornament.

Annabella's zeal as an educator stemmed from the unusually broad education she had herself received as the only daughter of wealthy, liberal, forward-thinking parents. She had studied history, poetry, literature, French, Italian, Latin, Greek, drawing, and dancing. Annabella was adamant that Ada would also benefit from a high-calibre education. Annabella was rich enough, and confident enough, to get what she wanted.

Yet while Ada was lucky in the education she received, she had scarcely more ground for optimism than any other intellectually enthusiastic woman of her day as regards finding an outlet for her mental energies after her education was completed. At a time when most human brainpower was comprehensively squandered, with intelligent working-class men as well as women obliged to live lives of drudgery and penury, toiling at monotonous work in filthy, noisy, and unhealthy conditions, Ada was fortunate only in belonging to a class that did not need to labour in such a fashion. Otherwise she had been born with the most severe of handicaps for anyone who wanted to live the life of the mind: she was female.

For a woman of Ada's day and social class who wished to lead a mentally fulfilling life, the opportunities were close to non-existent. There was generally little alternative but to marry, produce children, and live for one's husband. Annabella, to whom Ada was in thrall for much of her life, was conscious of how disastrous her own marriage had been. She was determined Ada would marry an aristocrat who could offer Ada a secure, comfortable domestic life. Ideally, Annabella wanted Ada to marry into the *older* aristocracy. There was a snobbish appeal at the time for titles that were more than a century old.

Annabella, rich, influential, and strong-willed, was used to having her wishes obeyed. Ada's yearning to lead a life of the mind, readily expressed even in the letters she wrote as a teenage girl, was thus doomed from the start. She was destined to spend much of her life aching to use her mind, but confronted with the day-to-day reality of children, nannies, servants, running a household, and dealing with a husband's whims. While not as oppressive and destructive of the intellectual life as the need to earn a living by spending twelve or more hours a day minding a dangerous and noisy machine that emitted clouds of dust, Ada's daily routine remained an endlessly frustrating one.

Faced with the reality of the limitations on a woman's role, middle-class or upper middle-class women confronted oppressive obstacles to making any real progress in advancing intellectual careers. Some, such as Jane Austen, the Brontës or—later in the century—George Eliot won careers for themselves through successful authorship. When they did, the reality of their struggle was likely to be a key subject of their books; Charlotte Brontë's *Jane Eyre* (1847), for example, is largely autobiographical in its account of the predicament of an intellectually gifted young woman forced to deal with the rigid limitations of life as a governess.

It was also true that, occasionally, enormous talent along with a stroke of good fortune might give a woman an opportunity to escape the bonds of domesticity. Mary Somerville, the close friend of Annabella and Ada who achieved international renown as a mathematician and scientist, first became interested in mathematics when she amused herself by solving the often challenging mathematical puzzles that were frequently published in Victorian embroidery magazines. The fact that such puzzles were published in such magazines at all is somehow curiously pathetic, as if the editors were all too aware of the wasted intellectual energies of their female readers and were trying to do what they could to give them something more weighty to think about than household chores and, well, embroidery.

Ada Byron met Charles Babbage for the first time on Wednesday 5 June 1833 at a party in London. He was one of many scientists there. Annabella noted in a letter to a friend that the party pleased Ada more 'than any assemblage of the grand monde'.

Babbage was forty-one years old at the time; Ada just seventeen. Babbage's wife Georgiana had been dead for six years. He had settled into an arduous, stressful, and often lonely bachelor life. It would be fascinating to know what he and Ada spoke about on their first encounter, but no account of the meeting or any relevant letters survive. We do know, however, that in a letter Ada wrote soon afterwards to her mother, she described her feelings for Babbage as a 'fondness' which is 'by no means inconsiderable'.

Their friendship ripened quickly. On Monday 17 June 1833 Ada and her mother went to Babbage's home. There, he demonstrated to them the working portion of the Difference Engine. This was the completed one-seventh of the machine he displayed at his soirées. It amounted to the pinnacle of his efforts to build the machine.

Ada's fascination with Babbage's ideas is clear from her letters. She found witnessing the operation of the Engine—or at least the part that Babbage had been able to complete—one of the most remarkable experiences in her life. She continued to be in touch with Babbage, and she and her mother often visited him when they were in London.

About eighteen months after Ada first met Babbage, she was with him, her mother, and Mary Somerville on the evening of Monday, 15 December 1834, when as we have seen Babbage used metaphors of poetic intensity to communicate to his guests his great excitement about his recent work on what turned out to be the Analytical Engine. The very fact that Ada became aware of Babbage's work on the Analytical Engine in this way shows she played no role in *originating* the idea of the machine. But geniuses need committed, visionary, and passionate disciples if they are to make the most of their ideas. Ada was precisely such a disciple, and Babbage was fortunate indeed she had decided to follow him.

By all accounts Babbage continued to see her fairly often during the next few months. Her mother frequently made a third at these meetings, but not always, and the correspondence between Ada and Babbage does not give any grounds for suggesting that Ada ever made use of a chaperone when she saw Babbage. One cannot in any case readily imagine someone as self-possessed and confident as Ada having much time for chaperones.

Ada and Babbage both had much to gain from their friendship. She would have been flattered by having the ear of one of the great intellectuals of her age; Babbage no doubt found her youth, beauty, and fascination with his work a gratifying diversion at a time when he was slaving away on his Engine without much tangible result.

The progress of their acquaintance can be followed in the surviving correspondence between them. This consists of eighty-five letters from Ada to Babbage and twenty-five letters from him to her. There were certainly many other letters that have been lost. The proof of this is that there are often references in the correspondence to other letters that have apparently not survived, while in 1853 Babbage wrote a letter to Lady Byron's solicitor in which Babbage referred to an 'extensive correspondence' he had carried on with Ada 'for years'. Indeed, even while I was researching this book, a small cache of letters from Ada to Babbage turned up in the storeroom of the Northumberland County Archive in the north of England. It is possible that more letters between them may yet materialize.

Did Babbage wonder whether Ada might possibly have made a wife for him? There is evidence that in the 1830s Babbage put out feelers among a few of his acquaintances to enquire whether they knew of any lady who might be a suitable new Mrs Babbage. Remarriage, at least at that time, was apparently not on his mind. True, Ada was much younger than he was, but such age disparities between the man and the woman were by no means rare in his day. Surviving letters from Ada to friends offer evidence that she did not find Babbage unattractive, and she might have imagined (not necessarily correctly) that marrying him would have fulfilled her intellectual longings.

But if Ada herself had ever speculated about the possibility of marrying Babbage she would have known only too well that Lady Byron would never have entertained it. Annabella was adamant Ada would marry a landed aristocrat of the best blue blood, not a man of science, however wealthy.

On 8 July 1835, after elaborate machinations by her mother, Ada married a Lord William King, a kind but not especially bright man who on the whole turned out to be a good and considerate husband.

William certainly fitted the bill when it came to blood of the right hue. His title was more than a century old; it dated from the early eighteenth century. Antique titles were much praised. Nor was Ada indifferent to William, who was good-looking in a fairly vapid way. Above all, he was rich, and had travelled enough in Europe to seem interesting, at least to start with. And then there was the physical side of marriage. Ada took to it with relish. Several letters she wrote to William shortly after their marriage make clear how much she enjoyed love-making.

On marrying, Ada became Augusta Ada King. In 1838, when her husband was elevated to the title of Earl of Lovelace by the young Queen Victoria, Ada became Countess of Lovelace. Today, she is generally referred to as Ada Lovelace.

How did Babbage feel about Ada's marriage? Annabella evidently thought he might have been jealous. She wrote Ada a letter asking 'but has Babbage cut you since your marriage?'— surely not something a mother is likely to write to her recently married daughter without cause. But there is no other evidence of Babbage's feelings about the matter than this extremely oblique reference.

It didn't take Ada long to realize that William, as well as being no intellectual, was not even very interesting. Rich as he was, there was always something slightly futile about him. To take just one example, he spent a great deal of time and money designing and ordering the construction of tunnels at his three country houses. The precise purpose of all these tunnels was never clear, and it is even possible there wasn't one.

Early in the marriage, Ada started to find her husband's lack of a proper occupation and any overall purpose intensely irritating. This was evidently a problem throughout their marriage; one of the letters recently discovered in the north of England and penned by Ada on Christmas Day 1846 amounts to a ticking-off for Babbage for, as she saw it, obstructing the procurement of a possible appointment for William. '*You* can have no conception of what my husband is, when his home *alone* occupies his irritable energies,' she writes. Ada was sure that an occupation, while certainly not necessary for William financially, would above all stop him hanging around the house and annoying her. She craved a husband who would do great things, *be* great, stride to fame and illustriousness with her by his side and understand her own pressing needs for an intellectual life. But William was not that man.

Yet Ada did what was expected of wives in those days: presented her husband with children. William and Ada's son, Byron was born on 12 May 1836; their daughter, Annabella, on 22 September 1837. A second son, Ralph, appeared on 2 July 1839.

Ada was often ill during these pregnancies. Her constitution was not strong; from the ages of thirteen to sixteen she had been partially paralysed and largely bedridden with an unidentified illness which may have been polio. Like most married ladies of her class, she found herself in charge of hiring and firing staff, balancing the books, and specifying recipes for meals as well as being a good, compliant, and pliant wife.

For the first few years of her marriage, Ada devoted herself to life as a wife and mother. After Ralph's birth, though, her thoughts returned with redoubled energy to her neglected mathematical studies. She became determined to find a distinguished mathematical and scientific tutor who would guide and accompany her on her intellectual quest.

Who better, she wondered, to fulfil this role than Babbage himself?

But Ada did not feel confident about approaching him directly. So, using what she thought was a cunning sleight of hand, she wrote to Babbage asking him to try to find a tutor for her. Her hopeful optimism shines through in the tone of the letter; there is little doubt she hoped he would offer to carry out the job himself. Like a cunning old carp, Babbage replied courteously to her letter and promised to keep his eyes open, but did not rise to the bait. And as things turned out, he not only failed to pick up her hint, but he didn't suggest anyone else, either.

In the summer of 1840 Annabella came to the rescue. She found a new tutor for her daughter: a well-known mathematician named Augustus De Morgan. Under his guidance Ada made rapid progress in studying her favourite subject. For the first time in her life she seems to have found some real intellectual fulfilment. Buoyed by the success of her work with her new tutor, she looked around for a new intellectual challenge.

Menabrea's paper on the Analytical Engine might have stayed as obscure as the learned Swiss journal in which it was published had not Ada decided that translating it into English would neatly achieve two objects she considered close to her heart. Firstly, it would give her the opportunity to publicize the important work being done by her close friend Babbage, of whom she was now seeing a good deal more than ever before. Secondly, the translation work would allow her to advance her dream of pursuing an intellectual career which would lift her above the demands of motherhood, running three homes, and looking after a wealthy but ineffectual husband.

Ada was conscious of the difficulty of her task, but was convinced she was more than equal to the job. She launched into it with characteristic energy. Her French was excellent, and her best writing has a fluency, clarity of expression, and mastery of metaphor and image that on occasion even recalls her father's fluent and expressive prose.

Here is a flavour of Ada's translation, from a section that discusses manual calculation machines (as opposed to Babbage's planned *automatic* machines).

> The chief drawback hitherto of most such machines is, that they require the continual intervention of a human agent to regulate their movements, and thence arises a source of errors; so that, if their use has not become general for large numerical calculations, it is because they have not in fact resolved the double problem which the question presents, that of correctness in the results, united with *economy* of time.

The paper then gets right down to the heart of the matter.

> Struck with similar reflections, Mr Babbage has devoted some years to the realisation of a gigantic idea. He proposed to himself nothing less than the construction of a machine capable of executing not merely arithmetical calculations, but even all those of analysis, if their laws are known. The imagination is at first astounded at the idea of such an undertaking; but the more calm reflection we bestow upon it, the less impossible does success appear, and it is felt that it may depend on the discovery of some principle so general, that if applied to machinery, the latter may be capable of mechanically translating the operations which may be indicated to it by algebraical notation.

It is interesting how expertly Ada gives her translation a 'feminine' feel that embodies an affectionate regard for the invention. Her tone is lighter, more expressive, and generally more emotionally committed than that of Menabrea's French. She achieves this effect partly by using words that have something of the fascinated, even gushing, schoolgirl about them: 'gigantic' is an example. Of course, this part of her work is in fact a translation and she is not at this stage writing her own material, but this same enthusiastic, emotional, well-informed tone is carried into her Notes.

How exactly did her own contribution—the additional Notes—come about? Babbage explains this in his autobiography. The passage in question is the only occasion in *Passages from the Life of a Philosopher* when he mentions Ada's name at all. The book is more an account of his scientific life than of his friendships or personal life. For example, he does not even mention his wife Georgiana *once*.

As Babbage says in *Passages*:

Some time after the appearance of his [Menabrea's] memoir on the subject [the Analytical Engine] in the Bibliothèque Universelle de Genève, the late Countess of Lovelace informed me that she had translated the memoir of Menabrea. I asked why she had not herself written an original paper on the subject with which she was intimately acquainted? To this Lady Lovelace replied that the thought had not occurred to her. I then suggested that she should add some notes to Menabrea's memoir: an idea which was immediately adopted.

We discussed together the various illustrations that might be introduced: I suggested several, but the selection was entirely her own. So also was the algebraic working out of the different problems, except, indeed, that relating to the numbers of Bernouilli [sic], which I had offered to do to save Lady Lovelace the trouble. This she sent back to me for an amendment, having detected a grave mistake which I had made in the process.

The notes of the Countess of Lovelace extend to about three times the length of the original memoir. Their author has entered fully into almost all the very difficult and abstract questions connected with the subject.

These two memoirs taken together furnish, to those who are capable of understanding the reasoning, a complete demonstration—*That the whole of the developments and operations of analysis are now capable of being executed by machinery.*

We see that Babbage is at pains here (the italics are his) to emphasize Ada's authorship of the Notes. He is even happy to concede that she corrected one of his mistakes. Bernoulli numbers are regarded as important by mathematicians because they allow an exponential series to be derived. Ada uses them in her Notes to illustrate how the Analytical Engine could be used.

Incidentally, Babbage was not quite right about the length of the Notes compared to the original translation: the Notes were, as we have seen, about twice as long.

Babbage's reference to Ada in *Passages* was supplemented by a paragraph Babbage wrote to Ada's son Byron on 14 June 1857, nearly five years after Ada's death, and seven years before *Passages* was published. In the letter Babbage observed to Byron Lovelace, 'In the memoir of Mr Menabrea and still more in the excellent Notes appended by your mother you will find the only comprehensive view of the powers of the Analytical Engine which the mathematicians of the world have yet expressed.'

There is no doubt about Ada's enthusiasm for Babbage's work. But how much of the writing that formed her Notes was really hers? After all, Babbage—not Ada—had invented the Analytical Engine, and it is known she worked closely with him when she was working on her Notes. The Notes are often extremely technical: could Ada really have written them all from scratch?

Asking this question, which considerations of historical accuracy oblige us to ask, in no way patronizes Ada or women scientists. Even a modern computer scientist would have difficulty understanding the Notes at many junctures. We also need to bear in mind that Babbage wrote his autobiography more than twenty years after the collaboration with Ada actually took place. Perhaps he had forgotten the extent to which he helped Ada with her Notes.

What other evidence is available to address this issue apart from what Babbage says in his autobiography?

There are two types of evidence that can usefully be considered. Firstly, there is the circumstantial evidence based on what can be known about the working relationship of Babbage and Ada. Secondly, there is the evidence of the content of the Notes themselves.

As far as the working relationship between Babbage and Ada is concerned, there is a distinct indication that Babbage was the *second* party in the creation of the Notes. Ada ran the project herself and it was *she* who solicited *Babbage's* help when she felt she needed it, not vice versa. Indeed, she was proprietorial about her work, and on more than one occasion told Babbage off when he made tentative efforts to suggest she present the content of the Notes in a slightly different way. But she did supply the drafts of her Notes to him for fact-checking, and she was prepared to defer to him when she got something factually wrong, as she did on occasion.

And the evidence of the content of the Notes themselves? This, taken along with what is known about Ada's mathematical abilities, tends to suggest that in the more technical content of the Notes, Babbage's input was very likely substantial. It is difficult not to admire Ada, but the simple fact is that judging from the main evidence—Ada's correspondence with her tutor, and also with Babbage himself—she was an *enthusiastic* amateur mathematician rather than an especially gifted one. To take just one example, here is a paragraph from a letter she wrote to Augustus De Morgan. It seems fair to use this evidence to assess her knowledge of mathematics.

Dear Mr De Morgan. I have I believe made some little progress towards the comprehension of the chapter on notation of functions, & I enclose you [sic] my demonstration of one of the exercises at the end of it ...

I do not know when I have been so tantalised by anything, & should be ashamed to say *how* much time I have spent upon it, in vain. These functional equations are complete will-o'-the-wisps to me. The moment I fancy I have really at last got

hold of something tangible & substantial, it all recedes further
and further & vanishes again into thin air.

Evidence such as this, and there are several similar examples
available, leads one to suspect that Ada may have been less good
at maths than she thought she was. It is certainly difficult to
believe that she originated the *technical* content of the Notes herself, especially as it is not just mathematics in itself that is being
expressed, but a very specialized form of mathematics relating to
the Analytical Engine. One hesitates to risk being unfair to Ada's
memory, but it is hard to believe that Babbage's account that
the entirety of the Notes was basically all Ada's work can be
accurate. He must have given her more help than he implies.

Yet the discursive element of the Notes, which is in fact a
much larger proportion of the Notes than the technical material,
is quite another matter.

Here, there really can be no doubt at all that Ada was the sole
author. Indeed, the discursive material benefits enormously from
her writing style, which is effusive, full of analogies, even on
occasion slightly schoolgirlish. Babbage could write wittily, but
his wit is never too far removed from pedantry, and he does not
have Ada's emotiveness and gusto.

A final clue to how Babbage regarded Ada can be seen in how
he signs off some of the letters he wrote to her during the time
when they were collaborating. For example, at the end of a letter
he wrote to her on 9 September 1843 in which he talks of his plans
to come to see her at her country house in Somerset, he says:

> Farewell my dear and much admired Interpreter
> Evermost truly yours
> C. Babbage

This is a very similar salutation to one he uses just three days later
in a shorter letter he writes to Ada from Dorset Street. He concludes this one with the phrase:

> Ever my fair Interpretress.

Babbage, it seems clear, while in no way seeing Ada as a part-ner—even a junior partner—in his intellectual labours, did regard her as someone who was interpreting his work for a wider audience. This seems to tally pretty accurately with what is revealed by circumstantial evidence about the nature of their collaboration and by the different nature of their writing.

Once we reach the conclusion that the discursive and emotional element of the Notes was indeed Ada's, we can fully appreciate one particularly important aspect of the Notes: her brilliant and inspired appreciation of the crucial conceptual links between the Jacquard loom and the Analytical Engine.

Ada's Notes provide abundant evidence that Babbage's adoption of the Jacquard loom control system was the aspect of the Analytical Engine that interested her most of all. Indeed, it seems very likely that it may have sparked her fascination with Babbage's most ambitious calculating Engine.

We must remember that Ada and her mother were both aware of the Jacquard loom and the punched-card system it employed *before* Babbage himself had the idea of employing the Jacquard cards in his Analytical Engine. In the summer of 1834, six months before the December evening when Babbage revealed his intellectual breakthrough to his three guests, Ada and her mother toured the industrial heartland of northern England. They saw various looms in action, and Lady Byron drew a picture of a punched card used in weaving ribbons.

The idea of the Analytical Engine as a kind of Jacquard loom that wove calculations had a deep and enduring appeal for Ada. When we look carefully at Babbage's writing style compared with Ada's, we are driven to the conclusion that he saw the world, and mechanisms, in a much more literal, factual, and indeed *analytical* way than she did. For Ada, inventing metaphors for understanding science was second nature. Babbage hardly ever did this.

(*right*) Lady Byron's sketch of a punched card used in ribbon manufacture.

July 17th — Saw the ribbon Manufactory at
Coventry. The Patterns, with the exception
of the very simple ones, are all formed
by stamped pieces of paste board —

Some of the holes are filled up, others are
left open — and thro' the latter packthreads
pass, which afterwards direct the silk
threads. — The highest wages 15s. a week.
The children employed in taking the
ends of the silk which would make the
surface of the ribbons uneven, work
from 6 in the morning to 7 in the
evening. ½ hour allowed for breakfast —
1 hour for dinner — ½ for supper — stand
in the morning, sit in the evening — go
to Sunday school generally — one had begun
as early as 7 years old. Trades' union not
considered alarming — but there had been
frequent Strikes. — It was Mr. Pepin's Manufactory.
The rules of a sick fund were written on one
of the walls. —

At Derby de la Youel, saw the Baths — and
Spar Manufactory — where turning Machines
are used — went on to Derby at night.
July 18. Saw China — Manufactory. — The
Infirmary — where I was interested by a woman

But the real point—and this explains why Ada's contribution to the idea of the Analytical Engine is so important—is that *the brilliance of the conception of the Analytical Engine requires both a scientific and emotive perception if it is to be fully understood and expressed.*

This, in fact, tallies neatly with an intriguing point Ada herself makes in a letter to Babbage at around this time, when she insists that her friendship with Babbage should be founded on 'imaginative roots' as well as on the more fundamental scientific aspects of their collaboration.

In the letter, dated 5 July 1843, Ada quotes a question Babbage had evidently asked her in a letter she received from him, and which has presumably been lost, for no letter from Babbage to Ada containing such a question survives in the extant correspondence. She goes on to make an important statement about the difference between how she thinks and how Babbage does.

'Why does my friend prefer *imaginary* roots for our friendship?'—Just because she happens to have some of that very imagination which *you* would deny her to possess, & therefore she enjoys a little *play* & *scope* for it now & then.

Ada sometimes made extravagant and unjustified claims about her scientific and mathematical understanding, but no one who reads her letters or the Notes could deny the quality of her imagination.

The entire metaphor Ada uses throughout the Notes to describe how the Analytical Engine works is founded in the operation of the Jacquard loom. Her writing makes the link between the Jacquard loom and the Analytical Engine about as explicit as it could possibly be. As she says:

The distinctive characteristic of the Analytical Engine, and that which has rendered it possible to endow mechanism with such extensive faculties as bid fair to make this engine the executive right-hand of abstract algebra, is the introduction into it of the principle which Jacquard devised for regulating, by means of punched cards, the most complicated patterns in the fabrication of brocaded stuffs. It is in this that the distinction between the two engines lies. Nothing of the sort exists in the Difference Engine. We may say most aptly that the Analytical Engine *weaves algebraical patterns* just as the Jacquard-loom weaves flowers and leaves. [Ada's italics.]

This last sentence summarizes with great precision, and a really delightful clarity and expressiveness of language, the entire nature of the connection between Babbage's work and Jacquard's loom. Ada often pursues this connection elsewhere in the Notes. Her Note F, for example, starts with a mention of 'a beautiful woven portrait of Jacquard, in the fabrication of which 24 000 cards were required'. This, of course, is the woven portrait which so fascinated Babbage and to which he was so devoted. One can easily imagine that Babbage and Ada were utterly intrigued by the woven portrait of Jacquard. How one yearns for some record of their conversations about it!

And now look at how, in Note A, Ada pays particular attention to the importance of the Jacquard loom control system in the operation of the Analytical Engine.

In ... what we may call the neutral or zero state of the engine, it is ready to receive at any moment, by means of cards constituting a portion of its mechanism (and applied on the principle of those used in the Jacquard-loom), the impress of whatever special function we may desire to develop or to tabulate.

Ada proceeds to define the purpose and use of the cards in precise terms. What she says about these cards is exactly what could be said about modern computer software.

These cards contain within themselves the law of development of the particular function that may be under consideration, and they compel the mechanism to act accordingly in a certain corresponding order.

Elaborating in the same Note on the distinction between the Difference Engine and the Analytical Engine and on why the Analytical Engine represents such a revolutionary breakthrough, she observes:

> The former engine [i.e. the Difference Engine] is in its nature strictly *arithmetical*, and the results it can arrive at lie within a very clearly defined and restricted range, while there is no finite line of demarcation which limits the powers of the Analytical Engine. These powers are coextensive with our knowledge of the laws of analysis itself, and need be bounded only by our acquaintance with the latter. Indeed we may consider the engine as the *material and mechanical representative* of analysis, and that our actual working powers in this department of human study will be enabled more effectually than heretofore to keep pace with our theoretical knowledge of its principles and laws, through the complete control which the engine gives us over the *executive manipulation* of algebraic and numerical symbols. [Ada's italics.]

The following passage, which enlarges on the role of Jacquard's cards in the Analytical Engine's operation, is another typical example of the clarity and focus of Ada's own thoughts on the Analytical Engine.

> The bounds of *arithmetic* were however outstepped the moment the idea of applying the cards had occurred; and the Analytical Engine does not occupy common ground with mere 'calculating machines.' It holds a position wholly its own; and the considerations it suggests are most interesting in their nature. In enabling mechanism to combine together *general* symbols, in successions of unlimited variety and extent, a uniting link is established between the operations of matter and the abstract mental processes of the *most abstract* branch of mathematical science. [Ada's italics.]

And Ada concludes:

A new, a vast, and a powerful language is developed for the future use of analysis, in which to wield its truths so that these may become of more speedy and accurate practical application for the purposes of mankind than the means hitherto in our possession have rendered possible. Thus not only the mental and the material, but the theoretical and the practical in the mathematical world, are brought into more intimate and effective connection with each other. We are not aware of it being on record that anything partaking in the nature of what is so well designated the Analytical Engine has been hitherto proposed, or even thought of, as a practical possibility any more than the idea of a thinking or of a reasoning machine.

If this is not a stunning and brilliant anticipation of the computer revolution that was to take place more than a century later, what is?

Ada was profoundly excited by her work on the Notes. On Sunday morning, 2 July 1843, while she was still working on the Notes, she wrote Babbage a letter from her country home at Ockham, Surrey. The letter contained the following intriguing paragraph:

I am reflecting much on the work & duties for you and the engine, which are to occupy me during the next two or three years I suppose; & have some excellent ideas on the subject.

The letter suggests that Ada regarded her work on Menabrea's paper, and the Notes she produced, as just the first stage of her new career as Babbage's interpreter. But in the absence of any other worthy papers on his work being published in a foreign language with which she was familiar, what else did she think she could do for him?

The answer would come in another letter—the longest she ever sent Babbage. She wrote it on Monday, 14 August 1843. Covering sixteen pages of her close handwriting and more than 2000 words long (the full text of this fascinating letter is contained in Appendix 2 to this book), the letter constituted nothing more or less than an offer to handle, from henceforth, what would be regarded today as the people management, political, and public relations aspects of Babbage's work on the Analytical Engine. Ada admired Babbage enormously but she was certain that the pedantic and undiplomatic aspects of his personality handicapped him when it came to advancing the cause of his Engines. She was perceptive enough to understand something Babbage never understood: that advancing his project required not only technical wizardry but also a profound skill at dealing with influential and sceptical people.

As Ada wrote in the letter:

> I want to know whether if I continue to work *on* & *about* your own great subject, you will undertake to abide wholly by the judgement of myself (or of any persons whom you may *now* please to name as referees, whenever we may differ), on *all practical* matters relating to *whatever can involve relations with any fellow-creature or fellow-creatures.*

But Ada never got the opportunity she craved.

At the top of the long letter that Ada sent him on 14 August and which is to be found in the Babbage papers, there appears a pencilled note in Babbage's handwriting that states:

> Tuesday 15 saw AAL this morning and refused all the conditions.

Babbage's own note really is as abrupt and dismissive as that. Babbage could, sadly, on occasion be selfish, stubborn, and ungenerous of spirit where his work was concerned. In that short note he combines all three vices. Despite his respect for Ada's ability to articulate and popularize the most important project of his life, he could never see her as anything more than an 'interpreter'.

Ada's reaction to his decision is not recorded, but I believe she would have been enormously disappointed by what was a silly and arrogant decision on Babbage's part. She would still have required his guidance and on occasion he would have had to check her youthful impetuosity, but if she had thrown her energy behind his project in the way she wished, who could tell what success the Analytical Engine project might have achieved?

Ada's translation of Menabrea's paper on the Analytical Engine was published in September 1843 in the third number of *Scientific Memoirs*. Entitled *Sketch of the Analytical Engine invented by Charles Babbage, Esq. (by L. F. Menabrea, with notes by Ada Lovelace)*, it was respectfully received by the scientific and mathematical community. But it did not cause the sensation Babbage no doubt hoped for, nor did it prove to be the springboard to a literary and scientific career for Ada. Yet the translation and Notes furnish a beguiling description of a nineteenth-century cogwheel computer. It is a pity that so few people had the prescience to notice its importance at the time. Posterity, however, has viewed Ada's work in an infinitely more significant light.

Ada's life after her collaboration with Babbage was mainly a tragic waste of talent. She only lived for nine more years. Within a few months of the collaboration with her friend she developed physical and mental health problems and started to take laudanum—a powerful combination of opium and brandy—whenever she felt unwell. By 1845 she was suffering much of the time from nervous exhaustion, general debilitation, and what were probably the first signs of the cancer that was to kill her.

Gradually her health deteriorated. Finally, much too late to be of any possible use, a diagnosis was made of uterine cancer. She died on 27 November 1852, after suffering frightful pain that was only partly relieved by the laudanum and by a new drug, chloroform. Lady Byron and Ada's husband William were at her bedside, but Lady Byron had quarrelled with Babbage, and he was not allowed to visit the house. Ada and Babbage remained good friends during her final illness. Lady Byron was bitterly jealous of the extent to which Ada had confided in Babbage about family and personal matters, when she could manage to get in touch with him.

A week after she died, Ada was laid to rest next to her father in the small church in the village of Hucknall Torkard, Nottinghamshire, close to Lord Byron's ancestral home of Newstead Abbey. The father and daughter—whose lives were almost exactly the same length—now lie side by side in a tomb that has been permanently sealed since 1929.

And Babbage? He continued to labour on his dream of cogwheel computers to the very end. He was, indeed, still working on his designs for the Analytical Engine when, after a short illness, he died on 18 October 1871, aged nearly eighty. In his last years he was plagued by headaches and the noise of urban life. He socialized little, spending much time alone in his London home, living among the ghosts of his dreams.

A precious but tragic insight into Babbage's forlorn later life was provided at a mathematical conference in July 1914 by the statesman and scientific adviser Lord Moulton. Recalling a visit he had made to Babbage many years earlier, apparently in the late 1860s, Moulton painted a dismal picture of the price the gods had extracted from Babbage for having bestowed on him a vision of a computer, without granting him the tools—technological, financial, and diplomatic—to make his dreams come true.

One of the sad memories of my life is a visit to the celebrated mathematician and inventor, Mr Babbage. He was far advanced in age, but his mind was still as vigorous as ever. He took me through his work-rooms. In the first room I saw the parts of the original Calculating Machine [i.e. the Difference Engine], which had been shown in an incomplete state many years before. I asked him about its present form.

'I have not finished it because in working at it I came on the idea of my Analytical Machine, which would do all that it was capable of doing and much more. Indeed, the idea was so

A portrait of Charles Babbage taken in London in 1860 when he was 68. This is probably the last portrait of him taken in his lifetime.

much simpler that it would have taken more work to complete the Calculating Machine than to design and construct the other in its entirety, so I turned my attention to the Analytical Machine.'

After a few minutes' talk we went into the next workroom, where he showed and explained to me the working of the elements of the Analytical Machine. I asked if I could see it.

'I have never completed it,' he said, 'because I hit upon an idea of doing the same thing by a different and far more effective method, and this rendered it useless to proceed on the old lines.' Then we went into the third room. There lay scattered bits of mechanism but I saw no trace of any working machine.

Very cautiously I approached the subject, and received the dreaded answer, 'It is not constructed yet, but I am working at it, and it will take less time to construct it altogether than it would have taken to complete the Analytical Machine from the stage in which I left it.'

I took leave of the old man with a heavy heart. When he died a few years later, not only had he constructed no machine, but the verdict of a jury of kind and sympathetic scientific men who were deputed to pronounce upon what he had left behind him, either in papers or mechanism, was that everything was too incomplete to be capable of being put to any useful purpose.

But the 'jury'—whoever they might have been—were wrong. As things turned out, Babbage *had* left behind enough plans and drawings for a complete, working version of one of his machines to be constructed by an epoch that was better equipped to understand the audacious brilliance of his vision. All he had really needed was access to an effective and efficient precision engineering industry: not because the technology of his own time was not up to the job of producing components to the requisite tolerances—it was—but because Babbage required a reliable source of thousands of identical cogwheels to be supplied relatively promptly, and at a reasonable cost.

As for Ada Lovelace's vision of a machine that could process and memorize calculations, algebraic patterns, and even all types of symbolic relationships as adeptly as the Jacquard loom could weave silk, that was a dream just waiting to come true.

·10·

A crisis with the American Census

Hollerith's pioneering punched card tabulating machines made it possible for the first time for government and business to process large amounts of information in an efficient, economic and timely way—to act on the basis of current facts, as one of his associates put it, before they became ancient history.

Geoffrey D. Austrian,
Herman Hollerith, Forgotten Giant of Information Processing, 1982

Herman Hollerith was a strange man.

He was instrumental in taking the story of Jacquard's Web into the twentieth century, but he appears to have had no sense of technological destiny. His work in pioneering the development of effective accounting machines based around the Jacquard loom makes him one of the founding fathers of the high-technology era, yet he would have been deeply embarrassed to have been described as such. He invented a completely new type of machine—a device he called a 'tabulator'—that can reasonably be described as the first successfully completed loom to weave information rather than fabric. This made the world his oyster, yet Hollerith was so blind to the commercial importance of his invention that he almost let his rivals reap these commercial benefits rather than himself.

He travelled the world to meet with foreign governments to whom he tried to sell his machines as an accounting tool for their national census, yet he detested travel and could not wait to get back home to be with his family. His inventive talents finally won him a large fortune, but he had no particular love of money and few pleasures except his family, food, and work. He was unquestionably a genius, but he had none of the technological vision of Charles Babbage or Ada Lovelace. If there is anybody who would have been astonished to hear that Hollerith's inventions led directly to the foundation of the computer giant IBM and to a global revolution in information technology, it would have been Hollerith himself.

A strange man, indeed. Yet a remarkably inventive pioneer whose story is an essential part of the tale of the Jacquard loom.

On Tuesday 24 October 1871 Charles Babbage was laid to rest in London's vast Kensal Green cemetery. Hardly anyone attended his funeral. As far as the world of science was concerned, Babbage had been history for many years before his death. The excited crowds that had once attended his wonderful soirées had long since melted away. The only carriage that followed the hearse contained the Duchess of Somerset, an elderly lady who had become a good friend of Babbage in his old age.

As the gravediggers tossed the cold gravel and earth over Babbage's coffin, consigning one of the greatest scientists of all time to darkness, a funeral guest with a visionary knowledge of his work might have been forgiven for thinking that it was not only Babbage who was being buried. The very notion of a special type of Jacquard loom that could weave information rather than silk was, one might have thought, also being sent into oblivion. After all, Jacquard, Ada Lovelace, and now Babbage himself were all dead, and the harsh, stern, materialistic world of 1871 was completely indifferent to Babbage's plans for his Difference Engine and Analytical Engine.

True, the Jacquard loom was by now being widely used around the world, but it was merely regarded as a superb machine for weaving images into silk, and nothing more than this. Anyone who had declared that the loom had laid the ground for a technological revolution which within a century would transform the world for ever, would have been laughed out of court. Assuming, of course, that anybody had bothered to listen to them.

But great ideas may acquire a life of their own as they travel on through the years, getting into all sorts of new company, and even acquiring new appearances no one could possibly have foreseen. To borrow a point made by George Orwell, the idea might even seem to change beyond all recognition, yet at heart, like any living thing, it will remain the same. Indeed, great inventors are rather like blacksmiths forging shoes to be worn by the winged horse Pegasus. The inventors start by setting out to solve a practical problem, then their solution soars off into uncharted skies whose scope they could not have imagined when they embarked on their work.

What is true of great inventors generally is especially true of Joseph-Marie Jacquard. The modest, industrious Frenchman never guessed that the automatic punched-card loom he had devised to weave mesmerisingly complex and stunningly beautiful patterns into silk would one day evolve into a tool whose breadth of possible applications was close to unlimited. Jacquard had invented a method to automate a complex process. But controlling the warp threads that enabled works of art to be woven into silk fabric barely even began to tap the potential of punched-card technology.

Tapping that potential required not a theory but a practical

Herman Hollerith as a young man.

application which worked. And in fact in 1871 the man fated to
pick up the baton of Jacquard's idea and continue the race was
already an eleven-year-old boy, his passion for machinery and
technology thoroughly evident. Thanks to Herman Hollerith, the
metamorphosis of the Jacquard loom into the most powerful tool
ever created lay just around the corner.

Herman Hollerith was born on 29 February 1860 in the town
of Buffalo, New York State. He was the son of German parents
who had come to America in 1848 to find a more prosperous and
stable life at a time when the states that formed a not yet united
Germany were going through continuous political turmoil. His
father, Johann Georg Hollerith, was killed in an accident when
Herman was nine and Herman's brother, George, six years older.
The boys were raised by their mother. Her family had for many

years been locksmiths, a profession second in precision only to clock-making.

From the age of twelve Herman attended the College of the City of New York. In 1875 he transferred to the Columbia School of Mines. This seems an unpromising name for an academic establishment, but in fact the School had a reputation throughout the United States for fostering leading-edge technical work. It was a popular choice among talented young men intending to follow careers in science and mechanical engineering.

Hollerith's academic career at the School of Mines was highly successful. He received perfect grades of 10.0 in descriptive geometry, graphics, surveying, and mechanical engineering. He also pursued a passionate interest in photography, carrying a camera almost everywhere. When he graduated in 1879 nobody doubted he would have a brilliant career.

One of the closest friendships Hollerith made at the School was with a Professor William Petit Trowbridge, head of the engineering department. Trowbridge was also a chief 'special agent' for the 1880 United States Census. This was planned to be the first US Census to focus as intensely on economic issues as on population data. The special agents were expert professionals charged with helping the US Government to make sense of the information the Census was expected to yield. Trowbridge had been given particular responsibility for analysing data relating to energy sources and mechanical engineering. It made sense to him to invite his most talented graduates to help him. As the most gifted of all, Hollerith was given the opportunity to play a particularly key role in the work.

Trowbridge decided to offer Hollerith the job of collecting statistics on steam and water power used in iron- and steel-making. Trowbridge had great faith in Hollerith's powers. He introduced Hollerith to Dr John Billings, another special Census agent. Billings, as impressed with Hollerith as Trowbridge was, asked the young man to create a table of life expectancies for different age groups. Billings had particular responsibility for the division of 'Vital Statistics' which, in those days, simply meant statistics related to living persons.

As with every project he ever undertook, Hollerith made every effort in completing these assignments to standards that were not only high but even obsessively so. Trowbridge was so pleased with Hollerith's work that once the Census had taken place he allowed the young man the rare privilege of publishing a report on his findings under his own name.

Hollerith's report investigated issues such as the relationship between fuel consumption and productivity and the number of workers employed in iron- and steel-making. Today, these hardly seem especially exciting issues to us, but back at the close of the nineteenth century, at a time when the United States was really starting to flex its industrial muscles, Hollerith's work represented a valuable study. Furthermore, some of his findings were in fact quite momentous. He showed, for example, that while the use of water power had stayed almost constant over the past decade, there had been an increase of more than three hundred per cent in the application of steam-power to steel-making as a result of the new Bessemer and open-hearth steelworks.

This kind of information about industrial trends and major developments had never before been available in the US. America had embarked on a remarkable curve of industrial development and population growth, but no tools had hitherto existed to gather the data that the government needed to monitor developments, control resources, and formulate policy. It was precisely this kind of data that the 1880 Census was designed to furnish.

The task of ploughing through the information generated by the 1880 Census was daunting. As for the US population, it was rising fast; the 1890 Census loomed on the horizon like a forbidding monster. It would have a terrifying appetite for research and statistics, but there were no tools available to feed that appetite apart from the hundreds of thousands of handwritten slips of paper that lay at the heart of the census-taking process.

The United States was a comparative latecomer in developing a requirement for large-scale information processing. In the 1830s, when Britain, France, and Germany were making great strides in industrial development, the American economy was still largely an agricultural one. It was only after the end of the Civil War in 1865 that United States commercial organizations started to achieve significant growth. With this growth came industrialization. As the nineteenth century drew to a close, this became a scale of industrialization that no other country in the world could remotely match.

Yet the United States was actually helped in its drive towards industrialization in the second half of the nineteenth century by the comparative lack of existing industrial and technological infrastructure. It was not handicapped by an existing infrastructure which, if—say—it had industrialized at the same time as Britain, would by now have been eighty years old.

In particular, when the United States Government and individual corporations started to developed significant needs for information processing in the 1880s and 1890s, they weren't hampered by the kind of manual information processing centres that abounded in Europe, where hundreds of clerks, working in cavern-like offices, kept voluminous ledger books by hand or made prodigious numbers of manual calculations. Instead, United States public and private sector organizations enjoyed the privilege of starting their information processing activities largely from scratch. They were perfectly positioned to make the most of any state-of-the-art information processing technology that was available.

This, combined with the United States having historically been blessed with many remarkably inventive technology pioneers, helps to explain how the US was able to establish and maintain a significant advantage in information processing over Europe at the end of the nineteenth century. This edge persisted throughout the twentieth century, and is even more pronounced today, in the twenty-first century, than it has ever been.

The prospect of the 1890 Census was placing the information processing requirements of the United States Government under great strain, but in fact the US Government's need to gather and process information more efficiently had been an increasingly acute problem since the end of the Civil War. Before the War, the only data collection and collation organization of any size was the Bureau of the Census, based in Washington DC. The directive to conduct an annual census had been issued by Congress in 1790 in order to determine the 'apportionment' of members of the House of Representatives. The first ever United States Census, held the same year, estimated the young country's population as 3.9 million. With a relatively small population to deal with, the information was processed by no more than a dozen clerks over several months.

The information gathered by the ten-yearly American Census continued to be handled in the same manual fashion throughout most of the nineteenth century. In 1840, when the US population had risen to 17.1 million, there were twenty-eight clerks employed by the Bureau of the Census. By 1860 Congress had recognized that the Census could achieve far more than simply record population statistics. For this Census, the Bureau employed 184 clerks to deal with a census of 31.4 million people and to collate additional information about personal characteristics such as age, ethnic group, and occupation. For the 1870 Census there were 438 clerks to record data for 39 million people, and the final report amounted to 3473 pages.

By 1880 the US population had increased to some fifty million, which was primarily why the Census held in that year had placed such an immense strain on the Bureau's information handling resources. To collate the results, a total of 1495 clerks were employed to run a manual system known as the 'tally'. After the Census forms had been completed and returned, each clerk had to transcribe the results from a particular district onto a grid containing a dozen columns and numerous rows. The columns and rows represented different parameters such as age, sex, ethnic group, occupation, the name of the state where the person lived, and any other information that the Government wanted to record. These filled-in grids would then be analysed in an entirely visual manner by other clerks and eventually form the substance of a Census report.

Needless to say, this system was tedious, expensive, slow, and horribly susceptible to errors, much as errors—those 'sunken rocks at sea'—had crept into the mathematical and scientific tables of the early nineteenth century.

The 1880 Census findings had taken seven years to collate in every level of detail required, although Hollerith himself was able to complete the statistics for his own work much sooner. The 1890 Census would be even more massive than the 1880 one. The population of the US was shooting up; between 1880 and 1900 it increased to seventy-six million, a gain of more than fifty per cent in twenty years. Even without the benefit of this hindsight, it was obvious to everyone involved with the Census that it would not be long before collating all the results of a particular Census would still be going on when the *following* one took place. The consequences of this situation were not difficult to imagine: the US Government was in danger of getting itself into a position where it would *never* be able to catch up with itself in trying to monitor its rapidly increasing number of citizens and their activities.

Almost as bad was the fact that the need to handle everything manually placed overwhelming restrictions on the breadth and detail of the information gathered as well as severely limiting the extent to which the findings could be analysed and processed. At the very time when the US Government needed to understand changes in the economy and population of the nation better than ever before, it faced the prospect of having to *limit* the scope of its enquiries rather than *extend* them.

The need to find a better way to handle the findings of the US Census was becoming desperate.

Fortunately, the Jacquard loom came to the rescue.

·11·

The first Jacquard looms that wove information

The punched card was the key. It supplied the means by which units of information could be processed once, rearranged in new combinations, and processed again, until every bit of useful information was extracted. Hollerith had taken the crucial step in the development of his tabulating system.

Geoffrey D. Austrian,
Herman Hollerith, Forgotten Giant of Information Processing, 1982

Herman Hollerith never publicly acknowledged the crucial role the Jacquard loom played in his work. Unlike Charles Babbage, Hollerith was not the kind of man to give credit where credit was due. As far as is known, he never made any public pronouncements on any subject. He never wrote an autobiography, nor did he leave a single word of writing that suggests any awareness on his part that he was carrying the sacred flame of technological destiny.

Nonetheless, we are justified in regarding the machines Hollerith created as a special kind of Jacquard loom that wove information rather than silk. In effect, Hollerith's devices went a long way towards fulfilling Ada Lovelace's inspired description of the Analytical Engine as a kind of Jacquard loom that 'weaves algebraical patterns just as the Jacquard-loom weaves flowers and leaves'.

Hollerith's tabulation machines were not as ambitious as the Analytical Engine, but they worked and they made commercial sense. Even more importantly, they paved the way for the creation of a new kind of machine—computers—which *would* offer the entire range of function and features that Babbage dreamt his Analytical Engine would provide.

The link between Jacquard's work to develop a revolutionary loom and Hollerith's creation of automatic accounting machines is clear and provable. It is a crucially important connection, and indicates that Herman Hollerith's work is just as important a junction point between the Jacquard loom and the modern computer as Charles Babbage's. Babbage laid the *conceptual* ground between the Jacquard loom and the computer, but it was Hollerith who made the connection a practical reality.

Hollerith knew all about the Jacquard loom. His brother-in-law, a businessman named Albert Meyer, was in the silk-weaving business in New York. Meyer had discussed the operation of the Jacquard loom in detail with Hollerith. Indeed, Meyer recognized Hollerith's engineering talents, and tried to steer his brother-in-law towards the textile industry. He did not manage to do this, but the lengthy conversations Meyer and Hollerith had about weaving and textiles left Hollerith in no doubt that the Jacquard loom had pioneered a crucially important new way of storing information. Hollerith also enjoyed watching the Jacquard looms in action in Meyer's factories. Observing an expert work at a Jacquard handloom is an exhilarating experience; Hollerith was doubtless as fascinated by seeing Jacquard looms operating in New York as Babbage had been at witnessing a Jacquard loom working in Lyons.

The recollections of a professor at Cornell University, Dr Walter F. Willcox, who worked at the Census Office in 1900, also show Hollerith's fascination with the Jacquard loom. Willcox recalled conversations with the special agent Dr John Billings that had taken place about twenty years earlier. As Willcox remembered, Billings said Hollerith had discussed in detail with him how it should be possible to build a machine that could *automatically* count the findings of the Census and *record* the results on punched cards just as the Jacquard loom stored information on punched cards for weaving a particular pattern or design. In fact, Hollerith never pretended to have invented punched cards himself. He knew that the concept was Jacquard's, not his. As his biographer Geoffrey Austrian points out, 'Hollerith is often credited with the invention of the punched card, called the Hollerith card for many years and later more familiarly known as the IBM card. Yet, significantly, he never claimed it for himself. His basic patents always encompassed the use of punched cards *in combination with* his machines.' [Austrian's italics.]

But in any event, the entire concept of Hollerith's machines is so closely related to the way the Jacquard loom stores the information necessary to complete a weaving that it is inconceivable Hollerith could have designed his machines without the inspiration of the Jacquard loom. It would certainly be gratifying if Hollerith had left behind any writing or notes that proved the matter conclusively, but in practice the circumstantial evidence for the link between the two technologies is overwhelming and decisive. The momentous idea that the key concept at the heart of the Jacquard loom could be transferred to weave information rather than fabric did not, after all, die with Charles Babbage. Instead, like the smuggled silkworm eggs that brought the secret of silk to Europe, it merely went into a state of suspended animation. Almost twenty years later, in New York City, it woke up and started to breathe again.

And it has gone on breathing ever since.

During the early 1880s, Herman Hollerith discussed in detail with his mentors Professor Trowbridge and Dr Billings the problem of finding a mechanical solution to the challenge of collecting and handling large amounts of Census information. Both special agents were fascinated by the idea, but ultimately it was Hollerith who embarked on practical experimentation directed at solving the problem.

By 1882 Hollerith had become convinced that the problem *could* be solved by mechanical means. He had also started thinking about what the machines that could solve it would look like and how they would work. In the meantime he had to feed himself, and unlike Babbage he had no family legacy to take care of this mundane but vital matter. This may have been an advantage to Hollerith; inventors driven by the necessity to earn money are often more focused and direct in their work and their approach than those for whom this is not so. Fortunately for Hollerith, in the same year 1882, the Massachusetts Institute of Technology (MIT) offered him a post as an instructor in mechanical engineering. Hollerith was only twenty-two; the job offer was immensely flattering. He accepted it immediately.

Hollerith, all his life a powerhouse of energy and ideas, threw himself into his new job with devotion and gusto. Among the subjects he taught were the science of hydraulic motors, the design of machines, steam engineering, descriptive geometry, metallurgy, materials science, and even blacksmithing. Mechanical engineering was, at the time, the king of sciences. Mechanical engineers were seen, not unreasonably, as magicians whose tricks rarely went wrong. They built machines to carry out a myriad different functions, they constructed steam engines to power locomotives, ships, pumps, and factory machinery. Some mechanical engineers were already devoting themselves to meeting the challenge of creating horseless carriages. Armed with their specialized knowledge of metallurgy, hydraulics, power transmission, gearing ratios, and other lore, they were busily engaged on fashioning a new world all around them.

Indeed, mechanical engineering was developing so rapidly as a science that even Hollerith was taxed to the limit by the need to remain up-to-date with what was going on in the field. Computer engineers of today are likely to find that if they have a two-week holiday they may miss a crucial new development in computing. Similarly, those wishing to keep abreast of mechanical engineering in the late nineteenth century had little choice but to keep working at the coalface where knowledge was being sledge-hammered out of the rock of ignorance. Yet despite the demands of his new job at MIT, Hollerith devoted his spare time to building various special machines for handling census returns.

Hollerith began his work by experimenting with methods he knew were already being used. These included strips of paper or individual cards on which details were noted by hand. But he was only too aware that the use of handwriting made the whole process extremely cumbersome. Besides, for Hollerith the purpose of a census was not merely to make a gigantic list of people's names and addresses, but rather to analyse their lives and work in comprehensive detail. Soon he decided to continue his experiments using one of the very newest technologies available—electricity.

The first patent he applied for in 1884 covered the invention of a machine that represented information by a double row of holes punched across the width of a roll of narrow paper. Hollerith had decided to start his experiments by using paper tape because this was popular at the time as a means for storing information to be transmitted by means of the device known as the automatic telegraph. The automatic telegraph relayed information at a distance using electric current. In fact, the use of punched paper tape in the automatic telegraph also appears to have derived from the inspiration furnished by the Jacquard loom.

In Hollerith's initial patent, the position of the holes indicated a person's age, ethnic origin, whether a person was male or female and native or foreign born, or any other information that the Government wanted to record. Hollerith's idea was that the roll of perforated paper would be passed over a metal drum that advanced the paper and allowed it to interface with a counting device. Metal pins, forming a grid pattern in their layout exactly like the pins that pressed against the punched cards used in the Jacquard loom, were pushed against the drum by a spring. Each time they encountered a hole in the paper strip, they made contact with the drum, completing an electric circuit. The completion of the circuit activated an electromagnet whose action registered a '1' on a counter provided for that hole.

The idea of having the presence or absence of a hole stand for a numerical quantity or a specific item of information would be absolutely fundamental to information processing in the years ahead. It was, again, exactly the idea at the heart of the Jacquard loom.

But Hollerith soon found that his concept of making use of a paper strip was not as practical as he had hoped. The paper tape turned out to be difficult to handle when it came to locating particular information. It also had an unfortunate tendency of snapping at critical moments. Abandoning the paper tape entirely, Hollerith adopted Jacquard's idea of using punched cards.

The use of a punched card as the fundamental unit of his information processing system was the key to the whole success of the entirely new kind of machines that Hollerith now started to build. The punched card allowed units of information to be processed once, rearranged in new combinations, and processed once again until all the relevant information was obtained. In effect, the punched card became the world's first-ever system for processing information using a standardized data input device. Ultimately, the basic principle behind Jacquard's loom, Babbage's Analytical Engine, and Hollerith's system of recording information about people was identical.

These pioneers all depended on using punched cards to record single, separate, units of information about something important they wanted to process or monitor, whether this was the specification of a picture to be woven into silk, the data and results of a calculation, or the details of a particular person's age, address, religion, country of birth, domicile, occupation, or—indeed—anything else that was the subject of scrutiny.

One significant difference, however, between how Jacquard and Babbage had conceived of their inventions and how Hollerith visualized his was that Hollerith's inventions, unlike the Jacquard loom and the Analytical Engine, were not stand-alone devices. Instead, Hollerith's vision encompassed an entire family of machines that would be deployed in unison. Today, this approach to using information technology is familiar to us all. Modern computers are routinely part of a system, linked to modems, printers, faxes, and—of course—each other. In fact, in the 1890s, the idea of a family of machines being used to complete a task was familiar enough from the manufacturing industry; Hollerith simply borrowed this approach for his own work.

The first element in Hollerith's system, understandably enough, was the *card-punch*: the device that actually punched the holes in the cards. In the early days Hollerith punched the holes himself using a stylus or other sharp object. But, he soon found this an unbearably tedious task, as well as hardly the most rewarding use of his time. Yet the job had to be completed with great application, because if the punched cards were to be an effective medium for storing information that could be processed by machines, the perforations needed to be in precisely the same place on each card to indicate the same answers to questions.

To carry out the task with the accuracy and speed required, Hollerith developed an ingenious machine based on the pantograph, a device invented in the early eighteenth century to copy a diagram on a larger scale. It consisted of a jointed adjustable parallelogram with tracing points at opposite corners. The picture on the following page shows how Hollerith's pantographic punch worked. It allowed the operator to create a 'mapping' or symbolic relationship between the guideplate in the foreground, which showed the operator where the hole should be, and the

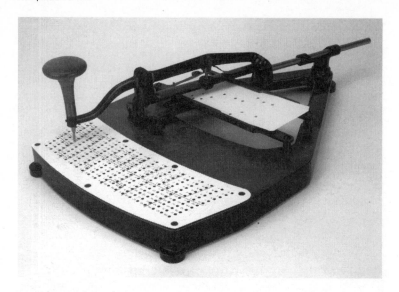

The pantograph punch.

actual card in the background. Provided that the operator indicated the right place on the guideplate, the pantograph punch would simultaneously punch the card in the right place. Using the pantograph punch, a clerk could punch many hundreds, or even thousands, of cards in a single day.

Once the punched cards had been prepared, the next stage was to count the occurrence of all the significant attributes of the person whose circumstances the card represented.

The ability of Hollerith's system to facilitate the counting of different elements of data contained on the punched cards was central to the successful operation of his system of handling numbers and other kinds of information by special machinery. The point is that the whole principle of Hollerith's system was based on the notion of one card being used per person. An

The Hollerith counter and sorter.

operator could employ the system to retrieve specific information about their ages. The cards would all be individually pressed against a grid containing blunt-ended needles. This grid was very similar indeed to the card-reading device featured in the Jacquard loom. In Hollerith's system, the needles were linked to dials which told the operator how many times the particular feature under examination had been triggered by the cards. The entire machine was known as the *counter*.

The counter worked in this way: when a needle ran through a hole in the card, it would dip into a little cup containing mercury. Mercury is an extremely good conductor of electricity. Mercury also has the remarkable property of being the only metal that is liquid at a comfortable room temperature. This makes it useful as an electrical contact where the contact needs to provide some flexibility as far as positioning is concerned; the flexibility that a liquid conductor can easily furnish.

In fact, there was no reason why the needle tip should not have made contact with any conductive metallic surface. But Hollerith was working at a time when electricity was not available from a main supply. Because of this, he needed to rely on lead-based accumulator batteries to operate his machines. These only provided a weak voltage. This made it essential to keep electrical resistance to a minimum.

Mercury provided the high-quality contact he needed, and does not readily corrode. Once the tip of the needle had dipped into the cup the circuit would be completed and the dial linked to that particular needle would advance by one increment. Pressing individual cards against the needles counted significant pieces of information gathered for the Census.

Like Babbage, Hollerith took care to build safeguards into his invention to avoid mistakes.

One problem that obviously had to be prevented was the danger of the punched card being entered into the system the wrong way up. To stop this occurring, every card was made with the same corner cut off to tally with a corresponding diagonally shaped place in the card slot. If the card was the wrong way round, the corner would not fit neatly into the card slot and the card could not be laid in its place at all. Hollerith also took steps to avoid obvious card-punching errors. He arranged for the system's wiring to deliver an audible alarm signal—usually a bell—to the operator if two mutually exclusive parameters (male and female, for example) were triggered simultaneously.

As well as this, it was useful for punched cards to be sorted according to the information they contained. The machine that did this was known, reasonably enough, as the *sorter*. Operators could sort cards into different boxes depending on the different criteria being investigated. If, say, the cards were being sorted according to the ages of the people each card represented, the different boxes in the sorter might contain cards representing the age ranges 0–10, 11–20, 21–30, and so on, with the final box containing cards for people older than 90. The card-sorting process was controlled by each dial being connected to a switch that opened the lid of a small box whenever the dial advanced by one increment. The operator who had pressed the card against the needle switches would see the lid of one box open and would drop the card into that box.

These processes of counting and sorting cards might seem very slow and cumbersome from the description given. However, operators could become surprisingly adept at working with this equipment. Skilled operatives could count and sort several thousand cards every day. All the same, we can be fairly certain none of these operatives would have wished their working day to be a moment longer than it was.

Like many inventors, Hollerith was never entirely satisfied with his current range of machines. He was continually refining and improving them. The pantographic punch was eventually superseded by keyboard-driven punches. Once these were invented there was not much more scope for improving the devices that actually punched the cards. Hollerith could never do away with the requirement to punch in the raw data. Someone needed to do this laborious data-entering activity. In practice, Hollerith's entire system relied on legions of card-punch operatives. He did, however, eventually develop an electric punch that needed only the lightest pressure on the key in order to activate it.

The fundamental principle that basic data needs to be entered manually still holds true in principle today. No matter how sophisticated a computer system might be, somebody, somewhere, has to enter into the system the raw information—numbers, words, or other variables—that are being harvested in the first place. Modern optical scanners take some of the burden away, but in practice the manual entering of data is still extensively required for most applications. What is true of data is also true of the often very large numbers of lines of computer programs. These lines are known within the computer industry as 'code'. A huge computer program writing industry has sprung up in India, for example, where so-called 'software factories' employ large numbers of talented Indian computer programmers who (whether rightly or wrongly) can be paid a good deal less than their counterparts in the United States. These programmers write many of the major programs—often requiring millions of lines of code—that power standard office and domestic desktop computers and laptops.

With little choice but to enter the data (i.e. punch the cards) manually, Hollerith saw that the real scope for improvement of his machines lay in the counter and the sorter. In particular, he realized that it should be possible to develop a counter that would automatically 'read' each card without an operator having to press it manually against the grid. Here again one sees an intimate relationship to Jacquard's conception of the punched cards being linked in a chain and momentarily locked and presented against the needle-grid automatically instead of being pushed in place by a manual operator.

Hollerith's efforts to improve his technology were intensified as a result of the competition he encountered from a rival in the design and manufacture of punched-card machines. This man, James Powers, was also a brilliant engineer who made many important innovations. The actual invention of the *automatic* sorter—whereby the sorting process took place without any human intervention at all—is generally ascribed to Powers rather than to Hollerith, although competition between the two men was so quick-fire and fierce that establishing priority between them over their most important inventions is far from easy. But in general, with some exceptions, Powers always lagged a little way behind Hollerith.

By around 1900, Hollerith had created a new system that made extensive use of automatic machines. The cards were first placed in a hopper that compressed them with a carefully weighted block to ensure that only one punched card was released from the hopper at a time. The individual cards were now capable of being processed faster than the eye could see. Each card was extracted automatically from the pack, pushed against tiny rods that sensed which holes were punched and which were not, and then dropped into one of various boxes depending on the results the tiny rods had found when pressed against the card. This process was basically analogous to the operation of the Jacquard loom; the difference was that the result of the cards being pressed against the mass of rods was that the cards themselves were sorted into different categories rather than that certain warp threads were raised to form a line of weaving.

Hollerith also introduced major improvements in the counting and sorting procedure. He had initially used clock-style dials to allow the operator to count the different numbers of cards in the different boxes, but soon found that it was easier for the operator to count accurately if the clock faces were replaced by cogwheels that allowed the user to read off the total number of features counted from a little 'window' in the machine. Eventually, as we saw above, the sorting machines did not even need manual operators to drop the cards into boxes. The cards moved over rollers, a sort of miniature conveyor belt, into whatever box had been triggered by the switches.

Hollerith felt that his new generation of automatic punched-card sorters and readers needed a new name. Around 1900 he started calling his automatic machines 'tabulators'. He hired the most talented engineers of his age to keep improving the pioneering innovations in what was rapidly turning into a completely new industry.

An important additional development in 1907 saw the tabulators being fitted with ingenious electromagnetic printing mechanisms so that the user would be presented with a printed document incorporating the totals from each operation. Fitted with printing mechanisms, automatic tabulators were transformed into enormously powerful automatic tools for information handling and analysis. Indeed, the tabulator went on to greater and greater success, with punched-card systems forming the basis of important and popular information handling resources that remained in use throughout the first six decades of the twentieth century. Further improvements in the machines greatly increased their speed, accuracy, and information handling capacity. Even as late as 1960, the data processing giant IBM was still earning more annual revenue from punched-card systems than it was from computers.

Today, one of the most curious aspects of the history of computing is how comprehensively the importance of tabulators has been forgotten. They were electromechanical digital machines that transformed the business of information handling in the first half of the twentieth century, yet today tabulators are only to be found in a few science museums around the world.

By the time Herman Hollerith went into business with his tabulators, the American economy was also being transformed by other ingenious office machines. Two of the most important were the typewriter and the cash register. The technology of these machines did not embody an enormously subtle and momentous idea as did the Jacquard loom, Babbage's Analytical Engine, and Hollerith's machines. All the same, the quality of their mechanical engineering was remarkable.

Hollerith's inventions belonged to the same new wave of business machines. Yet there was an important difference in the nature of the market for tabulators compared to the one for other machines. Typewriters and cash registers were useful in almost every type of business and in every area of industry and commerce. Hollerith's tabulation machines, on the other hand, at least at first, seemed to have only one application: census-taking. It was a precarious basis for a business. After all, the US Census, by definition, only came round once a decade.

There was another important difference, too. Mass-market business machines were manufactured and sold by highly focused organizations that emphasized sales success and the meeting of customer needs. Hollerith, on the other hand, operated in a far more informal and even academic way. He had fanatical commitment to the technical quality of his machines and immense talents as an engineer and inventor. All the same, Hollerith essentially ran his company as a research and development organization.

For Hollerith, commercial success came late. His company only became a global force in the business machine world when control passed to a more commercially minded entrepreneur. Between 1884, when Hollerith filed his first patent for a census tabulation machine, and 1911, when he sold the company, Hollerith built up a business that was remarkable more for the level of innovation it brought to its product development than for its commercial acumen and the dynamism of its sales force.

Another problem that handicapped Hollerith was that he tended to hire out his machines rather than selling them outright. He had little alternative: his customers were governments that only required the machines for the duration of the Census. Governments had no motivation to buy the machines for permanent use. At first, when the machines were still being perfected technically, renting them out was not entirely commercially satisfactory. Their complexity meant that they had to be repaired and fitted with replacement parts regularly, and Hollerith was responsible for this and had to swallow the cost.

Later, however, Hollerith found that hiring out machines that were technically reliable could, after all, be a good commercial ruse. Each machine might, over a period of years, earn back its purchase price many times. The fact that he tended in most cases to take rental machines back once they were obsolete and then melted them down helps to explain why few tabulators survive today.

Hollerith also insisted that his government customers purchase his blank punched cards, a reasonable enough insistence since punched cards could no more be re-used than could franked stamps. Some of Hollerith's customers tried to economize here by manufacturing their own cards or buying them from rival manufacturers, but the machines needed a high-quality card to run smoothly, and customers generally accepted the need to buy Hollerith's, which were of a better quality than any produced by his rivals.

Hollerith's first major success was the 1890 United States Census, when his machines were employed to automate the entire counting process. The 1890 Census, which had looked so forbidding when viewed from the perspective of 1880—when only manual means were available to handle the findings—was thoroughly tamed by the brilliance, efficiency, and practical advantages of Hollerith's first generation of machines. The whole event was a triumph for the perfectionist inventor; the machines worked so effectively that they made the previous, manual, means of handling a census look primeval. The US Government reckoned to have saved more than two years' work and $5 million of taxpayer's money, a prodigious sum by the standards of the time.

But just as important as the cost saving was the fact that the Hollerith machines made it possible to draw the widest range of information from the Census data. The most complicated tables showing relationships between different variables could be produced at no greater cost than the simplest. As for the accuracy and rapidity of the machines, everyone involved with the 1890 Census agreed that these were remarkable. As one admiring journalist wrote at the time: 'The apparatus works as unerringly as the mills of the Gods, but beats them hollow as to speed.'

Unfortunately for Hollerith, the very effectiveness of his machines eventually made them temporarily redundant, for by 1893 there was little left to do on the Census. The rented equipment was returned, leaving Hollerith with a warehouse full of dormant tabulation machines.

With no prospect of the US Government using the machines again until the Census of 1900—an assignment for which, in any case, he would have to submit a tender—Hollerith decided to seek new government markets abroad. This decision was an example of his short-sightedness as a businessman; he would have done better to have marketed his machines for commercial use back home. But Hollerith was overly focused on the role the machines could play in handling a census. Like Babbage, he could on occasion be limited in his horizons.

His short-sightedness in this respect was particularly surprising for quite apart from anything else, the decision to seek new markets abroad was an uncomfortable one for Hollerith personally. He hated travelling and had little or no interest in seeing foreign countries. Many of his letters home from abroad are written during times of great political excitement in the countries he visited. Another man might have relished this turmoil, even tried to take part in it as Jacquard had done. But Hollerith was an engineer, not an explorer. His letters simply reveal a world-weary and homesick inventor who can't wait to get home.

Foreign contracts also took longer to come to fruition than Hollerith had hoped. Much of his travel in the 1890s was done on a shoestring. At one point he received a cable from his wife telling him she did not have a cent in the house; he managed to wire her $20 to keep her going until he returned. But eventually he succeeded in persuading several foreign governments to rent his machines. The resulting contracts gave him financial security for the first time in his life.

Perhaps Hollerith's most notable success was convincing the Tsarist Russian Government to use his machines for Russia's 1896 Census. What would nowadays be described as the 'customer benefit' of the machines was too compelling to be ignored, even by the decadent and anachronistic Romanov Government. The Russians were as interested in benefiting from the cost savings and comprehensive analysis facilities offered by Hollerith's machines as the Americans had been. Not for the first time, technology had shown itself to be a fundamentally neutral kind of tool that readily crossed national borders because it met needs that were relevant to people everywhere.

On 3 December 1896, Hollerith organized his commercial ventures into a corporation he called the Tabulating Machine Company. Despite the success with foreign governments that he was now winning, he still refrained from developing his business in the private commercial sector. This appears to have been partly the result of his lifelong sense of being a researcher and inventor rather than a businessman, and also because his eyes were by now on the big prize of the 1900 United States Census.

After a nerve-racking tender—Hollerith had to compete against other inventors who had developed different types of tabulation machines—he did in fact win this extremely important contract. The Census sustained the Tabulation Machine Company for another three years, but when the work wound down, Hollerith was back where he had been ten years earlier.

It was only at this point that Hollerith, again faced with financial pressures and forced to confront the reality that his competitors were themselves targeting the commercial sector with great energy, launched a major initiative himself to win commercial customers in the United States. In doing so, Hollerith learned a lesson that all vendors of data processing devices and computers have learned at some point: that the biggest market for information processing systems is usually not the government sector, still less scientific or mathematical laboratories, but the offices of commercial organizations.

The commercial organizations of the nineteenth century could not, of course, dispense with the need for information processing if they did not have access to tabulators. They had to find a way of doing the job manually. To this end they employed armies of clerks, who worked with innumerable slips of paper at enormous expense and at a cripplingly slow pace. Fortunately for Hollerith, his competitors had not made too many inroads into the commercial market to prevent him from using his superior technology to become a prominent player within a few years. For the Tabulating Machine Company, it was a case of better late than never.

·12·
The birth of IBM

There will be IBM machines in use everywhere. The sun never sets on IBM.

Thomas Watson, IBM's founder, 1924

By the year 1911, Hollerith had more than one hundred major customers and hundreds of smaller ones. His rivals had moved into the commercial market before he did, but the superiority of his machines, his punched cards, his technology, and his quality control quickly allowed him to overtake them. Hollerith finally achieved dominance of the very market he had created: the market for what were becoming known as *business machines*, that is, machines which handled information. Alternatively, they might be seen simply as information-weaving Jacquard looms. Hollerith's Tabulating Machine Company was the first market leader in the US information technology industry.

It was true that the market itself was still not especially large. There were tens of thousands of corporations in the US that stubbornly clung to the old-fashioned manual ways of recording data on handwritten or typewritten slips of paper and storing them in vast libraries containing oceans of files. But the tide had turned, and the days when businesses could afford to deal with data in this manual fashion were numbered. The punched-card data storage, sorting, and retrieval system of Hollerith's tabulators was too powerful and effective to be ignored.

Hollerith was grudging about his success. There was a sense in which he never seemed entirely to believe in it. He accepted the financial rewards, but saw the growth and profits of his corporation as at heart irrelevant to his spirit of inventive enquiry and practical research. Still, he had finally been obliged to accept that the kind of research he loved to carry out needed to be funded by sales in the commercial world. He had also learnt from bitter experience that trying to base a secure business around processing census data was likely to end in debts and desperation. He had seen reason just in time.

But his health was not good. He was overweight. He had a lifelong addiction to good food and, according to his wife, a particular fondness for chicken salad. He was only fifty-one years old, but a lifetime of overwork and stress about his business was starting to take its toll. Hollerith's doctor, concerned about his patient's heart, was insistent that Hollerith change his lifestyle and play a less active role in running his business. Hollerith was not used to following the dictates of others, but on this occasion he found he had little choice.

In any event, there had been a development that might be seen as fortuitous. A few months earlier, a millionaire investor, Charles Ranlegh Flint, famous for his skill at putting together mergers, had approached Hollerith to see whether he had any interest in selling his business. Flint was the kind of businessman who turns everything he touches into gold. He had made several fortunes and kept them all. Short in stature, terrifyingly shrewd, and with the face of a thoughtful walrus, he found the tabulation machines fascinating. There was something about the eye-baffling speed of the machines, the way they processed hundreds of cards in a few seconds and sent each one into its own correct hopper after first counting it, that overwhelmed Flint's otherwise largely fiscal imagination. Watching a tabulator in action, Flint felt he was observing not just an ingenious machine but the very future of business.

Today, when working tabulation machines exist only in a few museums around the world, opportunities to see them operating are few and far between. But if you do manage to watch one in action it is not difficult to understand why Flint's imagination was so thoroughly captivated by the machines. They are indeed like a kind of magical loom weaving chaotic information into systematic, coherent, and logical patterns. Flint found the tabulators a perfect model for his own passion for imposing order on chaos and creating a pattern of understanding and direction from a maze of haphazard information. Crucial decisions in business are taken as much for emotional reasons as for logical ones, perhaps even more so. The tabulators inspired a level of emotion in Flint and in other powerful American businessmen that did indeed make these people feel certain they were witnessing a momentous breakthrough in humankind's mastery of information.

As indeed they were.

Hollerith had turned Flint down when the merger king made his first approach. But after a sobering appointment with his doctor, Hollerith thought again. He went to see Flint and re-opened discussions with him.

Flint's aim was straightforward enough: he wanted to combine Hollerith's corporation with other organizations to create a major force in precision business machines. Flint greatly admired the tabulators and what Hollerith had achieved with them, but he felt that the Tabulating Machine Company was too small to make the most of the opportunities out there that Flint was convinced were ripe for the picking.

The discussions between the two men progressed quickly. Finally, Hollerith made up his mind; he would sell his business to

Flint. The decision was helped by the fact that Flint had made Hollerith an offer he could hardly refuse—$1.2 million for his shares, about $20 million today. It was a vast sum for a business that was still in truth in a growth phase rather than in a state of maturity. One of the many reasons why Hollerith decided to accept Flint's offer was that many of his directors and employees, who had given him their lifelong loyalty, also owned shares in the Tabulating Machine Company and would become prosperous themselves as a result of the deal. Hollerith was not an especially emotionally expressive man, but he always remembered those who had shown him loyalty.

Once Flint had secured Hollerith's agreement, the merger king moved quickly. He made the Tabulating Machine Company the centrepiece of a merger with three other companies. These were the Computing Scale Company of Dayton, Ohio, which made scales and cheese slicers; the International Time Recording Company of Binghamton, which made clocks for keeping track of employees' hours; and the Bundy Manufacturing Company of Endicott, New York, which also made time recording machines. These other organizations were distinctly low-tech—even absurdly so—compared with the Tabulating Machine Company, but Flint was convinced the merger would work, and being wrong was not one of his vices. The new organization started to trade as a consolidated unit on 5 July 1911. Its name, drawn from its separate components, was now the Computing-Tabulating-Recording Company, or C-T-R. It was not an age when organizations needed trendy, eye-catching names in order to succeed.

As Flint himself later admitted, the merger was something of a first for him. His other deals had tended to bring together companies from the same industry horizontally, or merge customers with their suppliers vertically, or bring together firms involved in different steps of manufacturing or marketing: this was known as a circular merger. But the merger that had produced C-T-R was, as Flint put it when he looked back on it later in his career,

> neither horizontal nor vertical nor circular. In fact, it was so uncommon as to almost justify the description *sui generis*—in a class by itself.

Flint soon turned out to be right yet again. The C-T-R merger was a success from the outset. Flint was careful to ensure that a gospel of technical excellence and constant improvement of the new organization's products was fundamental to its business philosophy. Each area of C-T-R's activity drew energy from other areas in what would nowadays be termed synergy. Above all, Flint focused on meeting customer needs. Just as Hollerith's tabulating machines had been a response to the dire pressures facing those charged with processing the results of the 1890 Census, C-T-R's office machinery was designed to meet the current commercial crises in information handling.

As for Hollerith, he had sold his company and retired to the country. But he retained an interest in C-T-R, acting as a sort of 'elder statesman' in the organization, with a seat on the board of the former Tabulating Machine Company.

Hollerith might have remained with the company as a figurehead and board member for the rest of his life had he not come up against a formidable colleague at C-T-R. The man was Thomas John Watson, a human whirlwind who, while not in any sense a technological pioneer himself, had the vision and commercial acumen to see that it should be possible to transform Hollerith's machines into immensely powerful tools that would revolutionize how the world handled information.

Thomas Watson was born in 1874 in a small village, East Campbell, in upstate New York. His father was a lumber dealer. Watson began his career studying business at a local school of commerce before starting to work as a salesman in a retail store. From this he graduated to employment as a commercial traveller selling pianos and organs.

Later in life, Watson looked back at his time as a commercial traveller as one of his most important formative periods. He said it taught him some lessons he never forgot. He learnt that the role of the salesman is an honourable *profession* and that successful salesmen need to dress the part and behave like professionals. Another important lesson Watson said he had absorbed from his time on the road was that commercial salesmen were too prone to seeking solace in drink as a way of dealing with the loneliness and rootlessness of their jobs.

These guiding principles remained with Watson all his life. They explain his rigid insistence on dress code and his zero tolerance of drinking during working hours. Above all, Watson's early experience of working life convinced him that no commercial or industrial organization had any chance of lasting success unless it managed to recruit and hold on to energetic, focused, and highly motivated salesmen who were totally devoted to their customers.

In 1898 Watson joined the sales staff of the National Cash Register Company in Dayton, Ohio. A born salesman, Watson shot up through the corporate hierarchy. Within ten years he rose to the post of general sales manager of the company under the tutelage of its canny, ultra-competitive president, John H. Patterson.

Patterson played a vital role in defining the direction American business was to take throughout the twentieth century. Watson regarded Patterson as his mentor. In his own subsequent career he borrowed and built on many of the techniques Patterson had taught him.

In those days before mass-market advertising became a widespread feature of the business world, corporations were greatly dependent on their sales staff to drive home to the customer the reasons for buying a particular product or service. The trouble was, the very salesmen on whom a company's fortunes relied were frequently the least professional part of the corporate team, partly because of the rigours of the salesman lifestyle. There was not yet an infrastructure of comfortable, value-for-money business hotels where travelling salesmen could stay at the end of a long, hard day. Salesmen also usually felt neglected and even ignored by their employers. Small wonder that many became lonely and demoralized, paid too little attention to their customers, and spent what commission they managed to earn on drink.

Patterson's goal, which Watson subsequently eagerly adopted himself, was to head a revolution in how an organization saw its salesmen. Patterson believed that if he could achieve this within National Cash Registers, he could achieve a similarly revolutionary success in business. He set his company on a mission to, as he put it, 'exalt the salesman' and 'make a business-man out of him'.

Patterson led his revolution by introducing ideas that are nowadays taken for granted in business around the world but at the time were still new. Firstly, he took every step to motivate his sales force financially. His best salesmen earned regular raises and substantial cash bonuses. Patterson also offered salesmen status symbols that conferred on them the idea of belonging to an exclusive club. Top sellers won special gifts such as diamond tie-pins and gold-headed canes. He also brought a literal interpretation to this idea of expert salesmen belonging to a special club. Salesmen who achieved one hundred per cent of their sales quotas were invited to join what Patterson called the One Hundred Point club. Salesmen who achieved this status could feel proud to be among the fraternity of the most successful—and the best paid—businessmen in America.

Patterson sent his top salesmen to his own tailor and generally encouraged his salesmen to make themselves in his own image. Single-handedly he created an ethos of the salesman (and it almost always was a man in those days) as a glorious modern hero which has persisted in the business psyche of the United States, and most other capitalist countries, ever since.

Thomas Watson believed in these new approaches to motivating salesmen as fervently as Patterson did. By 1910 he had become Patterson's right-hand man. Then, as is so often the case when a pupil starts to approach a mentor in stature, mentor and pupil fell out. Patterson, having preached a gospel of helping his salesmen reach the very top, now paradoxically started to resent Watson's ability and success. It didn't help that Watson was enormously popular with the sales force; Patterson felt his own supremacy was being challenged. He and Watson had also become involved in a complicated and bitter antitrust lawsuit relating to NCR's attempt to control not only the market for new cash registers, but also for used ones. In this lawsuit Patterson and Watson were, in fact, on the same side, but the stress of the case and problems associated with it threw the final log onto the funeral pyre of their working relationship. One day in 1913, Patterson abruptly fired Watson.

Watson was stunned, but he never dwelt on negative matters for long and he recovered quickly. Besides, he was not exactly thrown into professional oblivion; he had already made a fortune at National Cash Registers and become a legend in American business by doing so. He was immediately offered a variety of attractive posts paying vast salaries.

To the astonishment of those who had worked with him and thought they knew him, he decided that the job which most appealed to him was Flint's offer of the post of general manager offered by the C-T-R company. Watson's acceptance of this job was not due to some superstitious preference for working for an organization whose name consisted of a three-letter acronym; he had, in fact, thought this particular career move out very carefully.

When Watson did take the job, C-T-R was still a tiny organization compared with NCR. But Watson had done his homework. He had investigated the nature of the patents C-T-R held. He understood, perhaps even more than Flint or Hollerith did, just how great the potential of these patents really was. At a more emotional level, Watson—like Flint—was fascinated by the intricacy, efficiency, and bewildering speed of automatic tabulation machines. Instinctively, he had a vision of the possibilities they offered for handling the information generated by America—and the world.

Watson always backed his own hunches. He negotiated a deal for himself whereby he would be paid only a modest annual basic salary, but also an annual commission of five per cent of the profits of C-T-R. If the organization grew according to an ambitious expansion plan that Watson had agreed with Flint, Watson could look forward to becoming the best-paid businessman on the planet. He insisted that as his future and C-T-R's future were now one and the same, everything would have to be done his way. Flint, not one to obstruct a star employee on the way up, agreed. The instant Watson assumed the captaincy of C-T-R, he started reorganizing it into the kind of company he wanted to lead to endlessly escalating success.

Abandoning Hollerith's low-key, almost embarrassingly unobtrusive sales techniques, Watson introduced the very same salesforce motivation practices he had helped Patterson deploy at NCR. Watson established a comprehensive system of sales territories, commissions, and quotas. Borrowing Patterson's idea of the One Hundred Point club, he founded what he termed, with typical increased directness, the One Hundred Per cent club. As with the One Hundred Point club, membership was not for life but had to be continually re-won every year by attaining one's own maximum sales quota.

Watson inspired his salesmen and paid big commissions for big sales. In return he expected a level of commitment and dedication unprecedented in American business life, even compared with other major organizations such as NCR. Working for Watson was infinitely more than a mere job; it was a vocation that ultimately took over one's personal life just as it consumed one's working hours. Anyone who did not see things that way had less chance of staying with C-T-R than a damaged punched card.

Above all, Watson demanded total loyalty to the corporation from everybody it employed. 'Joining a company,' he would tell his staff, 'is an act that calls for absolute loyalty in big matters and little ones.' His message was that the company was the employee's 'friend' and that a 'family spirit' combined with 'vision and faith', was as important for its success as an array of products that knocked the competition for six. He insisted that no drinking took place during business hours. His salesmen also had to adopt the same type of formal, neatly pressed suits and smart, pale shirts that Patterson had required his sales staff to wear at NCR. Members of the One Hundred Per cent club were celebrities within C-T-R, role models for everyone. When the club had a sales rally, this invariably took place at the best hotels in America. New York's Waldorf Astoria was a popular venue. The events had a jaunty and exciting sense of occasion and razzmatazz that made them more resemble a political rally than something run by a corporation. Salesmen and their wives were the company's celebrities and were showered with gifts and high-quality entertainment provided by America's leading artistes.

Nor did Watson only focus his attentions on C-T-R's sales force. One of the first things he did was to launch a research and development department dedicated to improving the tabulators. Typically, an activity that Hollerith had handled personally in a resourceful but ultimately informal and even undisciplined way was transformed by Watson into a formal, official, systematic function under his own direct control. Watson also very sensibly created a corporation-wide initiative to identify new types of markets where the tabulation system would prove irresistible. Again, Hollerith had run his business activities like an extended one-man band, but Watson was a 'big company' man who believed in everyone pulling together under a strong unified corporate branding, focus, and purpose.

One of Watson's many passionately held beliefs was that too many people failed to fulfil their potential because they didn't make enough effort to use their brains. He insisted that the word 'THINK' be posted on placards around C-T-R's offices and also on people's desks. Had personal computers been around in his day, no doubt Watson would have required that 'THINK' be placed on all the screensavers (as, indeed, it is on many screensavers at IBM today). Watson was a man who lived and breathed a mission of continuous business improvement, and he expected those he employed to do the same. It is an indication of his energy, drive, and determination that even *reading* an account of his career is exhausting. Watson ran C-T-R like a combination of international think-tank, army, church, political party (everyone certainly knew who the president was), and holiday camp. There were company flags, endless group photographs of everybody being cheerful together, a daily company newspaper, banquets at horseshoe tables, and even company songs. Watson pioneered what are today regarded somewhat erroneously as Japanese-style business practices decades before Japan actually got in on the act.

With the new man at the helm of C-T-R being about as much like Hollerith as a young tiger resembles an old bear, it was inevitable that before long Watson and Hollerith would rub each other up the wrong way. Both were strong-willed men with great minds that thought unalike.

Hollerith was at heart an academic inventor whose commercial success, though not his technological achievement, had in a sense been something of an accident. Watson believed that in business, as in life, things did not happen by accident but because you willed them to happen and took the practical steps to turn your wishes into reality. For Watson, making customers happy was the most serious thing in the world. Hollerith, though, was much more interested in technical issues. There were strict limits to his commercial intelligence, and he did not fully appreciate Watson's conviction that technical improvements are only important if customers gain increased levels of benefit from them.

Hollerith had a consulting contract with C-T-R. This left him free to do more or less what he wanted to do as far as involving himself with the Tabulating Machine Company was concerned. He came and went as he pleased. But Watson, needing total commitment from everyone in his team, could not accept that even the man who had invented tabulation machines should have this special status at the company. While always respectful to Hollerith, Watson saw him as one of the old guard who ought to accept that his main task now was to transfer his experience and knowledge into the minds of the new generation of employees. Hollerith had never been particularly keen on delegating expertise; his style was to try to do everything himself. Watson, on the other hand, was a motivator who saw that he could never grow an organization on the scale he wanted to create unless he could initiate legions of talented and hard-working people into his gospel of total commitment and total devotion.

To take one example: Hollerith regarded engineers as back-room boys who worked best when they were left alone. Watson, on the other hand, was quick to chase engineers out of the laboratory and into customers' offices to find out precisely what functions and features customers needed from their machines. Once the machines were developed and installed, Watson chased his engineers back into customers' premises again to deal with any problems customers might be having. Making the customer happy and eager to spend more was Watson's most fervently held credo.

The clash of Watson's and Hollerith's personalities was a classical example of a brash, energetic, visionary newcomer confronting a staid traditionalist. Watson wanted to create a new organization that would combine world-beating products with world-beating expertise at selling them. This was a different dimension compared to the formal, technically accomplished, but limited and hidebound ways that were habitual to Hollerith.

Such clashes are seen frequently in business. There is rarely an exception to the rule that it is the younger, more commercially visionary protagonist who wins. Not that it was much of a battle in the case of Watson and Hollerith. Hollerith was rich, tired, not especially healthy and past his prime. Watson was younger, hungry, and he had the world before him.

Faced with the choice between having to step in line and becoming a sort of senior acolyte of Watson or washing his hands of his whole involvement with C-T-R, Hollerith took the inevitable step. He resigned from the Board of the Tabulating Machine Company on 15 December 1914, two weeks after an ill-tempered discussion with Watson in which the younger man had made plain his intention to build a major sales organization out of C-T-R. Hollerith was, among other things, opposed to doing this. He had cautioned Watson that 'too many men are out for orders in this business ... regardless of the consequences'. But Watson did not care about the consequences. He wanted the orders.

Not unlike Jacquard himself, Hollerith spent the rest of his life in luxury and comfort, working farm land he had purchased, and gaining much enjoyment from the simple country life and the company of his family. In his old age he enjoyed sending his friends vast boxes containing produce he had grown on his farm. Sometimes the boxes contained forty or fifty pounds of vegetables.

Despite his differences with Watson, Hollerith continued to offer Watson advice. Watson magnanimously respected and valued this. He often invited Hollerith to discuss important issues with senior staff of C-T-R and on occasion to address factory-workers. No doubt Watson found it easier to do this now that Hollerith no longer threatened his power base. Hollerith gracefully continued to make suggestions for new types of machine. Watson invariably acted on these ideas, perfectly aware, with becoming modesty, that he himself would always remain a mere novice when it came to inventing revolutionary tabulation technology. He even arranged for a song about Hollerith to be written and sung at C-T-R social events. This, sung to the tune of the then-popular Laurel and Hardy song 'On the Trail of the Lonesome Pine' (which, rather bizarrely, was to enjoy another spell of popularity in 1975), went:

> Herman Hollerith is a man of honour
> What he has done is beyond compare
> To the wide world he has been the donor
> Of an invention very rare.
> His praises we all gladly sing.
> His results make him outclass a king.
> Facts from factors he has made a business.
> From the years good things to him bring.

Not a lyrical masterpiece, perhaps, but one could hardly deny that its heart was in the right place.

Watson, like any great leader who sees himself as surfing the leading wave of destiny, had a keen sense of history. He believed that in order to build for the future, an organization should have a sense of its own past. He urged Hollerith to save his letters and papers and wrote to him saying that there was surely a very interesting story to be told 'about the tabulating machine and the man who invented it'. He evidently hoped that Hollerith might tell this story, or at least find someone to whom he could tell it, but neither of these things happened. That had to wait.

In 1924, in recognition of the fact that the tabulating machines were making a far greater contribution to C-T-R's profits than any other product, Thomas Watson changed the company's name to International Business Machines Corporation Inc, or IBM for short. Another three-letter acronym, and one that has become perhaps the most famous commercial acronym in the world.

The intimate relationship between the Jacquard loom and Hollerith's tabulation machines, and the simple fact that these machines helped IBM develop into a global business, compel us to the irresistible conclusion that IBM had its origins in Jacquard's endeavours in Revolutionary France. And indeed IBM is, indeed, a direct descendant of the work that went on in

Thomas Watson and Charles Flint: two proud and successful capitalists standing to attention.

Jacquard's workshop during the last years of the eighteenth century and the first years of the nineteenth.

The world of people such as Charles Ranlegh Flint and Thomas Watson may, on the face it, seem a long way from that of Joseph-Marie Jacquard. But in fact the two worlds are connected by the clearest strand of logic. Any successful invention gives rise to improved versions of itself that will, in due course, foster the creation of an industry. Like religions, inventions tend to be created by lone geniuses but are typically developed and furthered by practical-minded, even ruthless, realists. The evolution of the Jacquard loom followed this pattern precisely. Few people were more practical-minded, ruthless, and realistic than Flint and Watson.

By 1928, with annual sales of about $20 million (worth about $300 million today), IBM was the fourth largest office machine supplier in the world. It still lagged behind Remington Rand, which made typewriters, and behind NCR and the Burroughs Adding Machine Company, but IBM was engaged on a process of dramatic growth, and was rapidly catching up with its bigger rivals.

Meanwhile, Herman Hollerith spent most of his time running his farm. On 17 November 1929, after only a short illness at the end of a retirement which had been almost entirely free of ill health, he died of heart failure. He was sixty-nine years old. He was already on the way to slipping from the consciousness of the commercial world. His machines, those extraordinary looms that wove information, were helping all types of organization— public and private—to win mastery over that very information.

As for IBM, it had now been founded, and the right person was at its helm. The stage was set for the Jacquard loom to start weaving the future.

·13·

The Thomas Watson
phenomenon

Most organisations, when they would dream of the exaltations
of the present, roll their eyes backward. The International
Business Machines Corporation has beheld no past so golden
as the present. The face of Providence is shining upon it, and
clouds are parted to make way for it. Marching onward as to
war, it has skirted the slough of depression and averted the
quicksands of false booms. Save for a few lulls that may be
described as breathing spells, its growth has been strong and
steady.

From a report in *Fortune* magazine, 1940

Herman Hollerith never liked the way he had been pressurized
by his first government customers to rent out his machines rather
than sell them. The governments preferred to do this because
they had no reason to own the machines permanently, but
Hollerith would have much preferred to have sold his machines
outright.

When Hollerith started making headway into the commercial sector, he again tried to sell his machines to his customers. But it turned out that commercial organizations, too, wanted to rent the tabulators. They knew how expert Hollerith was at improving his machines. They were confident that renting them from him rather than buying them outright meant they could always insist on being supplied with state-of-the-art machines. Meanwhile, Hollerith had little choice but to take the older machines back from his clients and melt them down.

By the time IBM was created, the system of renting tabulators to customers rather than selling them was taken for granted. During prosperous times it was, admittedly, by no means an ideal way of doing business: IBM would have much preferred its customers to have bought the older tabulators *and* the ones that superseded them. But when economic conditions worsened, as they did in the 1930s, the way Hollerith had been forced to do business not only saved IBM from likely disaster, but—really quite by chance—helped to catapult it to success.

The point was that even during the Depression, existing IBM customers, who would not have been able to afford to buy tabulators outright, found that they could still just about afford to rent them. The customers wanted to keep the machines if possible; they still needed them, and so the rental system was a blessing to them. As it was for IBM, which continued to win rental income from all its machines that were out with customers. There was also an attractive cost factor. The rental on an IBM tabulator repaid the machine's manufacturing costs in about three years. After this period virtually all the income from the machine was pure profit.

Another major income stream came from sales of blank cards that would be punched and processed by the tabulators. IBM enjoyed very much the same near-monopolistic control of the supply of these cards that Hollerith had enjoyed twenty years earlier. The cards had to be manufactured with great precision on special paper stock; most of IBM's competitors found it impossible to make blank cards to a similar quality of specification. Some, investing considerable sums, did manage to reach IBM's quality standard, but they did not enjoy the same economies of scale as IBM did and so could not compete with IBM on price.

The demand for punched cards remained high because once a card had been punched, it could not be re-punched to store other information. It could, of course, be re-used, but only as the repository of data already on it. IBM's clients had to punch new cards every single time they wanted to input new information. Many of IBM's biggest users required millions of cards every year. This was because new cards would usually be required not only for each new customer but even for each new transaction a new or existing customer made. Even modestly sized organizations would typically use tens or hundreds of thousands of cards annually; large organizations would indeed use millions.

The figures speak for themselves. During the 1930s, an otherwise wretched decade for capitalism, IBM sold about three billion cards a year in the United States alone. Revenue from cards accounted for only about ten per cent of IBM's annual income but approximately thirty-five per cent of its profit. It is amusing to record that one of the most technologically advanced organizations in the world at the time was deriving so much of its profit from selling cardboard cards, but—as Thomas Watson might have said—good business is where you find it.

A dangerous competitor to Hollerith and later IBM in the field of tabulation machine manufacture was, as we have seen, the engineer James Powers. In 1911 he formed his business interests into the Powers Accounting Machine Company. In 1927 this company was acquired by typewriter manufacturer Remington Rand, but Powers remained at the helm of the tabulation machine operation. The capital that Remington Rand brought to the party gave Powers an unprecedented level of strength to increase the scale of its competition with IBM.

During the 1930s both organizations fought vigorously against each other to offer the fastest and most reliable tabulation machines to the market. Yet IBM always had the better of the game. Its machines were known for being more reliable, more technologically advanced and better maintained than those sold by Powers. It was also widely recognized that IBM's salesmen were more devoted, dynamic, and deadly competitive than those employed by Powers and knew more about the machines they were selling. This technical expertise was substantially the result of Watson's creation of a special IBM training school at Endicott, New York.

During the early 1930s, IBM developed its '400' series of punched-card machines. IBM liked to brand its machines by numbers rather than names. At first the numbers had some connection with the number of prototypes that had gone into making the machine that was finally released to market. Eventually the numbers were merely used because they were a scientific-sounding branding. The numbering system was also convenient because different models in the same overall series could be numbered successively onwards. To give a more professional and systematic appearance, the numbering usually went up in steps of five or ten.

The most commercially successful machine of the 400-series was the model 405 Electric Accounting Machine, in its day by far the most advanced automatic tabulation machine in the world. IBM was soon manufacturing 1500 of these machines every year. Indeed, the 400-series actually continued to be manufactured late into the 1960s, when punched-card machines were finally superseded by computers.

With the advent of the 400-series machines, tabulators could count and sort punched cards at prodigious speeds. The fastest could handle 30 000 cards an hour, or about eight a second. This level of performance was continually improved until the final generation of punched-card machines of the 1960s was capable of handling about 50 000 cards every hour. The mechanical engineering that had delivered this level of performance was no less remarkable because it had stemmed from teams of highly trained engineers rather than from one gifted individual inventor.

By 1935 IBM had attained a dominating market share of eighty-five per cent of the US market for tabulation machines. Watson's organization had also established a significant presence in most economically developed foreign countries, and in several nations that we would nowadays refer to as developing countries. Thomas Watson's salary for the year 1934 was $364 432—worth at least $5 million at today's prices.

Nobody doubted that Watson deserved his success. With his vision of the need for technical excellence complementing his genius for inspiring and leading thousands of the best salesmen in America, he was not only running IBM but also inspiring millions of businesspeople throughout the world. Along with NCR's John Patterson, Watson played a key role in creating the cult of the salesman as the ambassador not only of a corporation and its products but also of a country.

What is particularly remarkable about Watson's success is that he achieved it during a decade of severe economic depression that had disastrous economic and political implications for so many other countries, particularly, of course, Germany.

The drastic downturn in the US economy that started with the Wall Street Crash of 1929 and which only really ended with the huge surge in production of World War II, posed the fundamental question of whether capitalism really worked at all. President Roosevelt's New Deal, implemented between 1933 and 1939, eventually helped to revitalize the economy by putting into practice what was at the time radical economic thinking. This abandoned *laissez-faire* economic attitudes and instead acknowledged that governments had a major role to play in invigorating a depressed economy by taking decisive and vigorous proactive measures.

In particular, the New Deal provided for the US Government to embark on public works programmes to reduce unemployment and to boost the country's infrastructure. Watson wholeheartedly supported Roosevelt and his New Deal. He had always believed in the company as a benevolent parent looking after its employees; for him it was a small step conceptually from 'big company' policies to 'big government' ones.

During the first five years of the 1930s, business machine manufacturers suffered an average decline in sales of about fifty per cent. IBM suffered along with the rest, but its punched-card sales and rental income provided it with a significant cushion. Despite the successful launch of the 400-series of machines, IBM had to be content with a significantly lower turnover while the national economy was depressed.

Responding to the crisis—one unprecedented in IBM's otherwise highly successful history—its Board of Directors recommended that the corporation drastically reduce its workforce to cut costs. The Board also recommended completely abandoning the continued production of tabulation machines while there were fewer customers to buy them.

But Watson, by now IBM's president, was the one in charge. He had carefully arranged matters so that his powers extended even to the point that he could veto a recommendation issued by the Board. He assured his fellow directors that there was bound to be an economic upturn sooner or later. Meanwhile, he insisted that the correct strategy was to continue to produce tabulators so that when the upturn came IBM would have the stock to meet the inevitably increased demand.

It was a fabulously audacious gamble, even given the fact that IBM had built up the financial muscle from its good years to be able to afford to take this level of risk. With hindsight, Watson's conviction looks wonderfully visionary and courageous. At the time, though, with IBM fauceting money every week on keeping its factories operating while revenue was dramatically reduced, it must have taken enormous courage for Watson to come to the decision he reached and to stick to it. After all, he had no evidence that there ever would be an upswing in the economy, apart from the general experience that economic events usually followed a cyclical pattern. With millions of Americans having to queue at soup kitchens and owning literally nothing but the clothes they stood up in, Watson's decision seemed to many like the act of a madman, or that of an emperor whose power had finally gone to his head. But Watson had the final say and the Board had no choice but to go along with him. And so IBM held on to its expensively trained workforce, building an enormous stockpile of machines in readiness for the upturn—assuming one came.

For IBM, this was its finest hour. During the worst economic crisis in America's history, Watson's confidence made him a hero of legendary proportions both within IBM and in the world beyond. Here was a man who practised what he preached, a business leader who had promised employees that their corporation would be loyal to them if they were loyal to it, and who had delivered on that promise even when he might have been expected to renege on it. Watson's decision also made him a hero with the general public, for he gave them hope that better times might be around the corner. And Watson acquired a heroic stature among America's politicians, who saw him as a beacon of hope during hopeless times. In recognition of his contribution to US business morale, Watson was voted to the presidency of the American Chamber of Commerce. He subsequently became an adviser and friend to President Roosevelt, one of the few men in the country of even higher calibre than Watson.

When the recovery did arrive, IBM's gamble paid off in spades. One of the many new legislative measures introduced as part of the New Deal was the Social Security Act of 1935. This laid the foundations for providing old age and widows' benefits, unemployment compensation, and disability insurance. In order to implement the Act, it was necessary for the US Federal Government to automate the employment records of the entire working population of the nation. IBM, with its stockpile of state-of-the-art tabulators and its factories that were already in full production, was perfectly placed to meet the Government's needs. It was a clear illustration of the principle that confidence in business, as in so many other things, often becomes a self-fulfilling prophecy. The question of the extent to which Watson's close government contacts helped IBM to win the contract has never been fully resolved, but here, as elsewhere in his career, Watson was ruthless in seeing an opportunity and exploiting it. If ever there was a man who believed that in life one creates one's own luck, that man was Thomas Watson.

In October 1936 the Federal Government took delivery from IBM of nearly 500 tabulation machines: a significant proportion of IBM's inventory. This enormous piece of government business represented a fabulous vindication of Thomas Watson's confidence. Between 1936 and 1940, IBM's sales in the US increased from $26.2 million to $46.2 million. During the same period, the number of people working for IBM in the US rose to close to 13 000. Information technology was becoming an enormous industry.

At this stage in their technological evolution, IBM's tabulators—those endlessly clicking, endlessly efficient Jacquard looms that wove information—were strictly electromechanical devices only. That is, they only made use of electrical circuitry and machinery. They did not use the technology of *electronics*, which would not come into prominence for another decade. But in some respects it is the very fact that the tabulators were purely electromechanical that made them so remarkable and wonderful to watch operating. The way they worked was much more obvious and transparent than is the case with the computers of today. Modern computers process raw data invisibly because everything happens at an electronic level.

But there was nothing invisible about the data processing carried out by IBM's tabulators. Weaving information from punched cards was what they did, and a tabulator's operation—

with the noise and the blurred speed of the cards as the card-pile shrunk in the hopper, as the cards were shot along conveyors to where in the tabulator they were needed, and as the printers fired out hundreds of results onto the paper rolls every minute—was a wonder of 1930s technology.

What a pity that Charles Babbage and Ada Lovelace never had a chance to see it.

·14·

Howard Aiken dreams of a computer

I have often admired the mystical way of Pythagoras, and the secret magic of numbers.

Sir Thomas Browne, *Religio Medici*, 1643

IBM's tabulators were busy handling more and more of the world's information, but Babbage's dream of giving the world a mechanical brain was still unfulfilled.

In fact, the need for a fast and reliable way of producing mathematical tables and performing a broad range of complex calculations with total accuracy was more pressing than ever. Even now, no reliable mechanical way of doing this automatically had been found, but a number of moderately effective calculation aids had been invented that took the edge off the difficulty while not solving it.

In fact, the calculation machine inventors who came after Babbage did not start out with his high ambitions to produce *automatic* calculators. Many computer historians, including the

Babbage expert Doron Swade, believe that Babbage's failure to produce a working example of his inventions rather queered the pitch, leading would-be pioneers to conclude that such a project was doomed to end in disappointment. And so, instead of trying to build automatic calculators, they aimed for the more modest objective of constructing strictly manual machines that would provide reliable results to arithmetical calculations too difficult to be done easily in one's head.

Back in the 1820s, a French businessman named Thomas de Colmar had invented a primitive version of a machine that would eventually turn out to be the first reasonably successful manual calculator. It became known in England as an 'arithmometer'. The arithmometer retained a low profile until the Great Exhibition of 1851, which brought it to the attention of a wider public.

The arithmometer was designed to assist with the kind of calculations necessary in offices. It was operated by 'dialling' the numbers that formed the calculation onto a set of cogwheels using a stylus and then turning a handle to carry out the calculation. In offices where calculations involving very large numbers were necessary—insurance companies and engineering firms for example—the arithmometer could be useful. But in the vast majority of offices, where book-keeping rarely involved very large numbers, the arithmometer did not provide much extra benefit. Book-keepers of the 1880s were routinely trained to add up or even multiply four-figure numbers in their heads, or with some use of scrap paper for arduous calculations. Most bookkeepers were adept at mental arithmetic and, being cheap to employ, their employers had no reason to look for a better solution. The result was that even in the 1870s, there was no other mechanical aid available than the arithmometer.

By the early 1880s, however, the continued pace of industrialization was prompting inventors to look again at the problem of finding a mechanical calculation aid. They knew that anyone who wanted to build a better calculator than the arithmometer would have to solve three crucial challenges. Firstly, there was the urgent need to speed up the rate at which numbers could be entered. Secondly, the speed of calculation had to be dramatically improved. Thirdly, if possible the machine needed to supply a permanent record of the result of the calculation. This last point was particularly important for banks, which were obliged by law to keep a permanent record of customer transactions.

For some inexplicable reason, many pioneers in automated calculation have curious names. Babbage was one perhaps. Another was Dorr E. Felt, a Chicago-based engineer who had the idea of building a key-driven calculating machine. Felt's innovation was probably inspired by the typewriter. The idea appears obvious now, but at the time it seemed an extremely ingenious and innovative development.

Felt used keys to enter specific numbers instead of laboriously entering them by turning a wheel or by pushing against a geared rack with a stylus. This approach offered two great advantages: speed and convenience. The keys of his machine—which he christened the 'Comptometer'—were arranged in columns, labelled from o to 9 in each column. Mechanically the Comptometer was relatively primitive. It necessitated ten keys (i.e. o to 9) for each column of integers. So, for example, a calculator designed to handle numbers up to 99 999 999 would need eight columns of ten keys.

Still, this machine was a big advance on any previous calculator. Yet its usefulness was limited. Its main function was to add and subtract; multiplications or divisions still had to be done manually by breaking down each component of the calculation into additions or subtractions. In 1886 a new company, Felt and Tarrant, started manufacturing the Comptometer on a production line. As the nineteenth century turned into the twentieth, Felt and Tarrant were selling 1000 Comptometers every year.

Gradually the pace of invention increased. By around 1910, the best key-driven calculators had become relatively sophisticated machines that used a combination of keys and gear-wheels to complete the calculation and in some cases even employed bells to indicate to the user when the calculation was complete. Generally the user was obliged to read off the result from a little 'window' in the machine and transcribe it. However, some calculators were fitted with paper rolls so that the result could be printed using a printing mechanism.

Despite advances in calculator manufacturing technology, calculators still used the cumbersome, slow method of repeated addition in order to carry out the all-important function of multiplication. The first calculation machine that could undertake multiplication and division properly instead of by repeated addition was known, optimistically perhaps, as the 'Millionaire'.

The Millionaire was about the size and weight of a small suitcase that had been filled to the brim with cogwheels. This, incidentally, was basically what it was. Its users, however, still belonged to a generation that was not especially mobile. For them, portability was not a major concern. The Millionaire sold well; by 1912 more than 2000 were in use. Two years later a motor-driven version became available. This model was also a success.

Mechanical calculators were put to use compiling a new generation of mathematical tables used to engineer all kinds of large-scale projects, including the great ocean liners of the first two decades of the twentieth century. For example, the marine engineers who built *Titanic*—at the time of its maiden voyage in 1912 the largest man-made object on the planet—relied on a variety of mechanical calculation aids at every stage of the work. Naturally, the great ship's unfortunate fate does not detract from the remarkable engineering achievement it represented.

Nor were mechanical calculators the only aid to calculation available in the early twentieth century. Hollerith tabulation machines, too, could be used for calculation purposes. Tabulators were by their very nature adding machines, and IBM eventually devoted considerable effort to modifying tabulators to maximize their usefulness for calculation purposes. By the 1930s it was routine for automatic tabulators to handle a wide range of calculations.

An additional resource that had become available during the late nineteenth century was the slide rule. This was still essentially a mechanical way of calculating, usually by moving two calibrated scales against each other. Slide rules were first invented in the 1860s, becoming popular about twenty years later. The best ones were able to carry out sophisticated arithmetical functions including multiplication, division, extraction of square roots, and even calculation of trigonometric functions and logarithms. But the slide rule's accuracy was restricted due to the problem of parallax (meaning that a precise reading varies according to the eye's position in relation to the calibration) and also because the accuracy of the result depended on the precision of the slide rule's calibration. There was, and is, always a limit to how precisely slide rules can be calibrated.

Despite the progress that had been made, Charles Babbage's ambition still remained well out of reach even during the 1930s. The problem was that if Babbage's dream of *automatic* calculation were to come true, a new kind of technology would have to come into play. In a similar way, for example, the invention of hot-air balloons did not make it possible for people to travel to the Moon; that had to wait until the rocket engine was perfected. Even by 1940, a device had yet to be invented that could undertake any type of mathematical calculation a user might reasonably want to compute.

In particular, there was no machine available to carry out tasks needing extensive and prolonged calculation. Some calculations could, in theory, have been enacted by clerks using state-of-the-art arithmetical calculating aids, but it would have taken many weeks for some of the more complex calculations to be completed, and many calculations could never have been realistically handled manually at all, even if years could have been devoted to the task, because the calculations were simply too demanding.

During the 1920s and 1930s there was increasing and justified concern among scientists, mathematicians, engineers, and many others that the lack of reliable automatic calculation machines presented a serious obstacle to continued progress in all the sciences, and of course also in mathematics itself. The concerns expressed in the 1820s about the inaccuracy of mathematical tables had returned, and with renewed force. Yet still there was no machine available that could placate them.

The first person in the twentieth century to try to solve the problem was a man called Howard Aiken. Aiken is as important a character in the story of Jacquard's Web as Joseph-Marie Jacquard himself, Charles Babbage, or Herman Hollerith. Aiken's work provides the final connection between the Jacquard loom and the modern computer.

Howard Hathaway Aiken was born in Hoboken, New Jersey on 8 March 1900. His father, Daniel H. Aiken, came from a wealthy and well-established Indiana family. His mother, Margaret, was a child of German immigrants. Howard was their only child. When he was still a boy, he moved with his parents and maternal grandparents to Indianapolis, the capital of Indiana and about 150 miles south of Chicago.

Life was difficult for the family. Daniel Aiken was an alcoholic and would often beat his wife. During one such episode, young Howard—already big and strong at the age of twelve (he would eventually reach six foot four)—grabbed a poker from the fireplace and drove his father out of the house. The family never saw him again.

Unfortunately, yet perhaps not surprisingly, Daniel's wealthy relatives responded to what Howard had done by declaring that they would have nothing more to do with either Howard or his mother. For much of his adult life, Howard worked to support his mother and his maternal grandparents. None of the relatives he supported seemed to think of actually working to support *him*. The possibility of Margaret Aiken engaging in remunerative work did not arise. In those days a middle-class 'lady' was usually unskilled and could not simply go out and find work. More to the point, she would have considered it an appalling social disgrace to have done so.

Aiken left school when he was still a teenager so that he could continue to provide for his mother and her parents. He found a job installing telephones. Later in life he enjoyed telling friends that he had installed all the telephones in the red-light district of Indianapolis. Despite the full-time commitment of his job, he did not abandon his education, but started taking correspondence courses in the subjects he enjoyed. One of his former school teachers went to see Margaret Aiken and pleaded with her to let him return to school full-time, but the family's financial circumstances meant that this was impossible. The teacher sought another solution. He found Aiken a job as an electrician's assistant for the Indianapolis Light and Heat Company.

In his new job Aiken worked a night shift so he would be able to attend school during the day. Probably few ordinary students would have been able to cope with such pressure, but Howard was very far from being ordinary in any respect. He not only graduated from high school, but continued his correspondence courses, too.

Throughout his career Aiken had a particular gift for winning the help and support of teachers and officials who got to know him. By all accounts he was an inspiring student, and others wanted to do their best for him. This talent, which stayed with him throughout his student days and beyond, played an important part in his success.

One of the officials who came to have the highest opinion of Aiken was the local superintendent of public instruction, one Milo Stewart. Because Aiken's need to work in paid employment had left him short of some of the school credits he needed in order to go to university, Stewart created a special examination so that Aiken could get these credits.

Stewart also wrote to every Midwestern public utility in a university town asking it to employ the young man. Aiken was eventually offered a job as night-time telephone operator by the Madison Gas and Electric Company, based in Madison, Wisconsin. It was because of this job offer that he chose to go to the University of Wisconsin. Aiken was twenty when he moved to Madison with his mother. He enrolled in an electrical engineering course at the University of Wisconsin, working at his new job from four o'clock in the afternoon until midnight and attending university during the day.

Edward Bennett, the chairman of the University's electrical engineering department, had a huge influence on the young Aiken. A quarter of a century later, in a letter Aiken wrote to Bennett inviting him to the ceremony marking the dedication of the world's first electromechanical computer, Aiken wrote, 'I sincerely hope it will be possible for you to be with us, for in a large part, successful completion of this machine was due to the careful preparation which I had as your student.' He signed the letter 'Respectfully yours'.

In a lecture he was to deliver in 1955 both in Sweden and in Germany, Aiken referred to the importance of Bennett's teachings in the shaping of his own intellect. Aiken recalled that Bennett had taught him how the development of any new body of knowledge passed through four stages.

The first stage, Aiken said, was *observation*, when the investigator knows almost nothing about the subject and so can do little else than make observations from nature and observe new facts. To take the study of electricity and magnetism as an example to illustrate Aiken's point, one might say that the study of these subjects had been in the observation stage throughout most of history until Michael Faraday started his important and hugely influential work in the early nineteenth century.

The second stage was *classification*, when the observer has sufficient facts to be able to separate them, place them under major headings, and rank the facts in the order of their import-ance. The study of electricity and magnetism engineering had reached the classification stage by about 1880, following major discoveries made by Faraday and other pioneers.

The third stage, *deduction*, was when a new science is born. By the start of the twentieth century the study of electricity and magnetism had become a science of its own: *electrical engineering.* Electrical engineering was a thrilling science to be studying at the start of the twentieth century, as much an exciting field of exploration as physical exploration by sea was for mariners in the fifteenth and sixteenth centuries. Electricity had yielded up many of its secrets. The more research there was, the more it became clear that the potential for applying electricity in everyday life was fabulously exciting and possibly close to unlimited. Furthermore, the speed of electrical processing was creating the possibility that new, powerful types of machines and communi-cation devices might be made which had the potential to revo-lutionize technology in every aspect of life. In Bennett's terminology, the deduction stage was when it became possible to deduce new facts from old. Knowledge would have attained what was, in effect, a critical mass and this fertile bed of knowl-edge would have created a new discipline in which all sorts of new ideas and initiatives could flourish.

Finally, there was the *methodological* stage, in which there is actually a danger of the findings and pronouncements being accepted without question, with original thought being stifled by the sheer need to master a vast body of knowledge before any original thinking could take place. A case could at least tenta-tively be made that computer science is close to reaching that stage today. That said, constant problems with bugs (i.e. errors in software) bedevil even mass-manufactured, highly popular pro-grams. This problem, and other problems such as the challenge posed by viruses and the often tortuous difficulty of integrating

213

disparate computer systems without causing failures in communication and conflicts between programs, indicate that in practice computing is a rather less precise science than might be expected.

In 1923 Aiken was awarded a Bachelor of Science degree in electrical engineering. He continued his studies at Wisconsin and later at the University of Chicago, working full-time as usual. This explains why he did not matriculate as a graduate student in Chicago until 1932. In practice his paid work complemented his academic career because his employers recognized his talents and gave him extensive responsibilities that provided practical training in the very subjects he was studying. For example, during Aiken's period as a graduate student, his jobs included that of electrical engineer responsible for the design and reconstruction of an electric generating station and also chief engineer of another company. At the time when Aiken was studying electrical engineering, the link between the new science and useful and major practical applications of it were particularly strong: Aiken's work helped him to be familiar with these applications. It was typical of the strength and resilience of his character that he turned an apparent disadvantage, such as the need to work full-time, into an advantage in this way.

Aiken continued his studies by enrolling at Harvard in the autumn of 1933. He did so well that he was invited to teach there for the academic year 1935–6. For several years he continued at Harvard, teaching and researching in electrical engineering.

From early in his career Aiken was known for his skill at visualizing abstract mathematical or physical situations in terms of physical processes. He was always deeply interested in any piece of mathematical equipment that allowed a mathematical process to be carried out by physical means. In particular, he was intrigued by the punched-card tabulation machines available in the 1930s and in mechanical analogue devices such as tide predictors. These were basically machines that used a complex combination of gear-wheels and gear-shafts to provide a reasonably accurate representation of the constantly varying action of the tides. The more advanced of these machines were used to handle complex differential equations. They became known as *differential analysers*. Differential analysers were wonderfully subtle pieces of machinery, but the fact of their being machinery placed inherent limitations on their speed, reliability, and accuracy. Eventually, in the age of electronics, the functions formerly carried out by tide predictors and differential analysers were performed by special types of computers known as *analogue computers* because they deal with information as a continuously variable quantity.

The vast majority of computers in the world today, however, are known as *digital computers* because they deal with information in the form of digits or similar discrete elements. Similarly, the Jacquard loom handled the elements of a complex woven picture in an essentially digital fashion.

Toward the end of his life Aiken traced his passionate interest in digital computers to his time at Harvard. He said, 'I was obliged to conclude that the area of electronics in which I was interested could never be explored properly with the calculating machines that were available at the time. The reason was that it would simply be too laborious and time-consuming to undertake the very numerous, lengthy, and extremely repetitive calculations needed to complete this exploration.' He believed that some way had to be found to ensure that the labour of calculating 'could be mechanized and programmed', and that an individual needed to have access to a very special type of device in order to do this properly.

More than a century had elapsed since Babbage had started work on his Analytical Engine. In 1936, the word 'computer' still meant in popular parlance a human clerk who actually carried out the mathematical calculations with pen and paper.

At the time Aiken was working on his doctoral thesis, most computers were women. The job was badly paid; men usually considered it beneath them. In practice, anyone undertaking a complex mathematical calculation in 1936 was just as dependent on mathematical tables as Babbage and his colleagues had been. The tables—laboriously put together by people working with calculating machines—were more reliable than those available to Charles Babbage and John Herschel, but they were still by no means error-free.

There were at the time only a few facilities in the United States capable of large-scale calculation. One of the best-known was the Computation Center at Columbia University, later known as Columbia's Watson Scientific Computing Laboratory, named after Thomas Watson himself, who sponsored it. By the mid-1930s the Computation Center was making extensive use of the latest generation of IBM punched-card machines that had been modified for calculation. The Center was engaged in important work involving the computation of lunar orbits; again, one is reminded of Babbage and Herschel's work for the Royal Astronomical Society back in the 1820s. But even with the latest punched-card machines, the work was hampered by precisely the same problem that had confronted Babbage and Herschel, and which Aiken also needed to solve in order to make progress in the field of electronics. The problem was that computers did not yet exist.

It was in 1936, while working on his thesis, that Aiken started to put together informal plans for a completely new kind of calculation machine. One of his mathematician friends recalls a discussion around this time, in which Aiken discussed the possibility of directing the activities of scores of computing units, initially talking in terms of racks of interrelated machines, each roughly equivalent to the mechanical calculators of the day. By late in 1936, the friend recalls, Aiken had started to give careful thought to how such a machine might have instructions inputted into it. He was determined to use paper-tape or cards but initially he had not decided whether the holes in the tape or cards should be sensed mechanically or electrically.

Finally, after careful thought, Aiken decided that he preferred the electrical way of sensing the holes in the tape or cards. In effect, Aiken had decided what Charles Babbage may well have decided had he been born in 1891 rather than a century earlier.

By April 1937 Aiken had made sufficient progress in his thinking about the general design of the machine, and the tasks it could perform, to start seeking support from industry for actually building the machine. Never one to waste time, on 22 April 1937 he presented his plans to the Monroe Calculating Company, America's largest manufacturer of calculators. He was given a warm reception by George C. Chase, Monroe's director of research. Chase subsequently published an account of the visit, during which Aiken explained that:

> certain branches of science had reached a barrier that could not be passed until means could be found to solve mathematical problems too large to be undertaken with the then-known computational equipment.

The plan that Aiken discussed with Chase amounted to an enormously important step on the way to the modern computer. As Chase recalled, it consisted of a proposal for a machine that

> provided automatic computation in the four rules of arithmetic; facilitated storage and memory of installed or computed values; established sequence control that could automatically respond to computed results or symbols, and provided a printed record of all that transpires within the machine; and a recording of all the computed results.

Chase was convinced that the machine specified in the proposal represented one of the most remarkable innovations in the history of technology. He did everything in his power to convince his colleagues at Monroe that the organization should build it. Chase accepted that the machine would be enormously expensive, but he tried to persuade the corporation that the machine would play a crucial role in every aspect of the company's development in the future. Chase evidently based his case on emphasizing how the machine would be an immensely important promotional tool for demonstrating to the world the sophistication of Monroe's technology. In essence he saw the whole thing as a brilliant public relations exercise.

The problem, however, was that the Monroe Calculating Company made calculators, but Chase was asking it to fund a completely speculative project for another purpose. The Board congratulated Aiken on the inventiveness of his ideas and the ambition of his plans, but declined to be involved. Its reasons were the inevitable huge cost of the machine, and a belief that Monroe was not the organization to build it. Chase's embarrassment at the rejection prompted a decision that would change the history of technology. He introduced Aiken to the one other corporation in the United States with the technical expertise, vision, experience, and—above all, money—to fund the project. IBM.

·15·

IBM and the Harvard Mark 1

If, unwarned by my example, any man shall undertake and shall succeed in really constructing an Engine embodying in itself the whole of the executive department of mathematical analysis upon different principles or by simpler mechanical means, I have no fear of leaving my reputation in his charge, for he alone will be fully able to appreciate the nature of my efforts and the value of their results.

Charles Babbage,
Passages from the Life of a Philosopher, 1864

Thomas Watson was a businessman rather than a technologist, but his intuition told him that his cherished organization could only flourish if it remained at what would nowadays be called the leading-edge of technological development. As things turned out, Watson's belief in this principle resulted in IBM being the perfect incubator for Howard Aiken's ambitious plans.

Charles Babbage, dreaming of mechanical computers in the early nineteenth century, had had little alternative but to try to build his machines in his own workshop. Howard Aiken was fortunate in being alive at a time when a successful organization with considerable experience in information technology existed and had money available to fund speculative research. As Aiken, with typical generosity of spirit, was subsequently to remark:

> If Babbage had lived seventy-five years later, I would have been out of a job.

Chase gave Aiken an introduction to a Professor Theodore (Ted) Brown who taught at Harvard Business School. It was a happy choice. An applied mathematician and a trained astronomer with a PhD from Yale in celestial mechanics, Brown was already a consultant for IBM and a friend of Thomas Watson. One of Watson's many talents was a gift for winning over people who were smarter than he was and giving them key positions at IBM. For example, Brown ran special training courses at IBM and often lectured staff himself. At the time he met Aiken, Brown was a member of the Advisory Board of a computation laboratory that Watson had set up at Columbia University. This meant he was directly and intimately involved with the problems of scientific calculation by machine, and well aware of the need to make progress in this area. Aiken met Watson through Brown.

Watson quickly took to Aiken. At the outset everything went very well between the two men. Watson was always intensely respectful of the kind of Ivy League academia that Aiken exemplified. Besides, there was a strongly philanthropic side to Watson's personality. He was committed to using some of IBM's resources for educational purposes and for the advancement of science generally, and Aiken's project was exactly the kind of venture that appealed to him. Watson arranged for Aiken to be introduced to IBM's senior engineer, a man named James Wares Bryce. Following this introduction, Aiken prepared a formal proposal for the new machine.

The proposal was entitled, with the directness that appealed as much to Watson as it did to Aiken, *Proposed Automatic Calculating*

Machine. Occupying twenty-three double-spaced pages, it was, in effect, the first blueprint of the computer age.

The proposal opens by discussing 'the desire to economize time and mental effort in arithmetical computations; and to eliminate human liability to error'. This desire, Aiken says, is 'probably as old as the science of arithmetic itself'.

After brief references to various early and unsatisfactory attempts to mechanize calculation, including the devices developed by Pascal and Leibniz, Aiken discusses the slide rule before turning to the work of Charles Babbage.

It is important to emphasize that Aiken did not have access to the detailed technical knowledge of Babbage's machines we have today. This knowledge has only been available since the 1970s, when painstaking research among Babbage's original papers unearthed it. Aiken's main source of information appears to have been Babbage's autobiography, *Passages from the Life of a Philosopher*. While extremely illuminating on the matters of Babbage's personality and outlook, this book is weak on the technical features of his two Engines. It would have been quite impossible to have built a Babbage machine from the material in it, for example. Still, Aiken would at least have learned from *Passages* that there were two basic types of Babbage machine— the Difference Engine and the Analytical Engine. He was also aware that the Analytical Engine (as he put it)

> pointed the way ... to the punched-card-type of calculating machine ... it was intended to use perforated cards for its control, similar to those used in the Jacquard loom.

Which is, of course, exactly the point.

In the proposal, Aiken moves on to discuss Hollerith's invention of punched-card 'tabulating, counting, sorting and arithmetical machinery'. Aiken takes pains to emphasize the historical continuity of the ideas that originate with Babbage. As he explains, the ideas then proceed through the work of Herman Hollerith and result in the proposal itself. The link he suggests between

Babbage's work and Hollerith's is misleading, however. There is no evidence that Hollerith knew of Babbage's ambitions. The real link is from Hollerith back to the work of Jacquard.

Aiken ends the retrospective part of the proposal by observing that the machines 'manufactured by the International Business Machines Company' have made it possible to achieve daily in the accounting offices of industrial enterprises around the world everything that Babbage wished to achieve. The somewhat fairly crude flattery here may, perhaps, be excused on the grounds that Aiken was doing everything he could to give his proposal the best chance of succeeding when presented to IBM. Aiken was planting in the reader's mind the notion that his new type of machine might also have important commercial applications in the offices of industrial enterprises.

Of course, Aiken's comment was in any case an exaggeration; IBM's tabulators, highly sophisticated as they were, were ultimately merely extremely fast card-readers and card-counters. They were not computers in the modern sense of the word, and it *was* a modern computer (technically a digital computer) that Babbage had been trying to build. And so was Aiken.

The historical material Aiken provides is an important link which in effect summarizes much of the story of Jacquard's Web. It furnishes a clear and unequivocal series of links that connect the Jacquard loom to the modern computer via Charles Babbage and Herman Hollerith.

Aiken's proposal states the case for constructing a machine capable of providing 'more powerful calculating methods in the mathematical and physical sciences'. He emphasizes that 'many recent scientific developments' are based on highly complex phenomena that cannot be properly studied until some means is found to make the abstruse calculations that would be necessary in order to track the phenomena, understand them, and predict them. There are problems beyond our capacity to solve, he writes, 'not because of theoretical difficulty, but because of insufficient means of mechanical computation'. This is essentially a twentieth-century version of Charles Babbage's lament at the lack of a reliable way of calculating mathematical tables.

Aiken specifies four design features that differentiate ordinary punched-card accounting machinery from the kind of machine he has in mind. The brilliance of his overall conception is shown nowhere better than here, where he lays down what is nothing less than a specification for the modern computer. This specification, along with the discursive material that precedes it, gives Aiken's proposal a claim to being one of the most important documents in the intellectual history of mankind.

The first essential design feature, Aiken points out, is that the machine must be able to handle both positive and negative quantities, whereas the punched-card machines could only deal with positive numbers.

The second design feature is that the machine must be able to handle all types of complex mathematical functions such as trigonometric functions. These describe different relationships between the angles and the length of the sides of a triangle and have always been of crucial importance in calculations relating to navigation and engineering. Aiken emphasizes that the machine must be a computing device in every sense of the word; that it needs to be able to do much more than basic addition, subtraction, multiplication, and division.

Thirdly, he stipulates that when the machine is carrying out a mathematical calculation, the calculation process must be 'fully automatic in its operation once a process is established'. This was a vital matter, very likely referring directly to Babbage's Analytical Engine.

Fourthly, Aiken specifies that the machine must be able to handle calculations that involve adding up figures horizontally in rows rather than only vertically in columns, which was all that the tabulation machines could do. Aiken pointed out that this feature was especially important for differential equations, where the computation of a value frequently depended on existing values.

Aiken goes on to assert that the adoption of these four features is all that is required to convert existing punched-card calculating machines ('such as those manufactured by the International Business Machines Company') into a new type of machine that would be, as he puts it, 'specially adapted for scientific purposes'. Aiken concedes that the number of components would have to be increased compared with existing machines. Yet it seems clear from the proposal that he envisaged the machine being constructed from the same components that IBM was already using. In fact, as things turned out, the world's first computer not only needed to be built from new types of components but also had to be operated in a manner radically different from even the most sophisticated automatic tabulation machine.

The proposal also specifies sixteen mathematical operations that would need to be built into the machine. These include the four fundamentals of addition, subtraction, multiplication, and division, logarithms and antilogarithms of base ten and other bases, as well as trigonometric functions. Like Babbage, Aiken emphasizes the importance of being able to *print out* the results so as to avoid the danger of human error in copying the results from a 'window' in the machine. The proposal ends with a general statement of confidence that, for IBM, the job of building the machine is merely a further step along a road already trodden:

> Suffice it to say that all the operations ... can be accomplished by these existing machines [i.e. IBM's latest generation of automatic tabulation machines] when equipped with suitable controls, and assembled in sufficient number. The whole problem of design of an automatic calculating machine suitable for mathematical operations is thus reduced to a problem of suitable control design, and even this problem has been solved for simple arithmetical operations.

Aiken delivered his proposal for the world's first computer to James Wares Bryce in early November 1937. Thanks to the enthusiastic support of Bryce, Thomas Watson, and other senior people at IBM, work soon got underway. At Bryce's suggestion, Aiken attended an IBM training school, where his existing detailed knowledge of IBM machines was brought completely up-to-date and where he gained important experience in actually using the machines.

Then, early in 1938, Aiken visited one of IBM's principal research and development laboratories at Endicott in New York. There, he started to work on the machine in conjunction with a team of Bryce's most trusted and able senior engineers. Bryce placed in charge of the project an extremely capable male engineer with the unusual name (yet another one) of Clair D. Lake. Lake was one of IBM's longest-serving engineer inventors. He had designed the first automatic printing tabulator in 1919, a formidable feat that involved linking, with great precision, the rapid punched-card counting mechanism with a series of electromagnetic relays—that is, on/off switches—which controlled magnetic hammer-type print heads. Aiken could hardly have asked for a better project leader, nor for a better team of engineers.

Needless to say, engineering of this calibre cost serious money, much more than Aiken had envisaged. IBM's initial estimate of $15 000 to build the machine was quickly raised to $100 000 (about $1.5 million at modern prices) once the full extent of the true complexity of the task was understood by IBM's engineers.

Early in 1939, IBM's board gave its formal approval for the construction of what was by now called an 'Automatic Computing Plant'. Once the project had been authorized, and the engineering team had started its work, Aiken's role became primarily that of a consultant and adviser rather than a full-time member of the engineering team itself. This was not due to any lack of enthusiasm on his part, but because during the early years of the work his knowledge of the technical issues was not extensive enough to allow him to assemble the machine itself. Later he had no time to work on the machine because during the war years he had full-time responsibilities within the armed forces.

During the war, Aiken's project became just one of several special projects that IBM was carrying out for the US Army. It was inevitable that it should become a military project in a time of war, although it was not immediately clear what practical military applications the machine would have. The most likely application would be calculations to do with ballistics and battlefield information. As these contain numerous variables, they could not be handled by IBM's most complex punched-card machines, let alone by individual clerks.

In the meantime, Aiken had joined the United States Naval Reserve, where he was assigned to the Naval Mine Warfare School in Yorktown, Virginia. With the rank of Lieutenant Commander, he acted as a senior instructor to young officers on the subject of electricity and electronics. Aiken appears to have enjoyed being in the Navy, but he was disappointed at being deprived by the war of the chance to be intimately involved with the work at IBM. By this point the actual construction of the machine was mainly a matter of building the components and assembling them, and IBM's own engineers were more than competent to do this. Aiken's moment of glory, however, was to come.

The war, coupled with the technical difficulties of the project, caused many delays in the schedule. IBM was not ready to run its first test problem on the computer until January 1943. It was a secret operation, further evidence that the US military was hoping the machine would offer useful war-time applications.

The complete machine was a wonderful piece of engineering. Fifty-one feet long, eight feet high, and two feet wide, it weighed more than five tons. Altogether it contained more than 750 000 parts and hundreds of miles of wiring. In operation, the

Howard Aiken as Lieutenant Commander.

machine was described by one commentator as making the sound of 'a roomful of ladies knitting'. All the basic calculating units were lined up and driven by a fifty-foot shaft powered by a five-horsepower electric motor (not steam, though). The machine could store up to seventy-two numbers in its memory. It was capable of three additions or subtractions a second. A complex multiplication took six seconds, while calculating a logarithmic or a trigonometrical function required about a minute.

The machine is today regarded as the first automatic digital calculating device ever to be successfully constructed. As such, it deserves to be regarded as the world's first computer. It was programmed by a sequence of instructions coded on punched paper tape. The word 'program', meaning a set of instructions for a mechanical device, was entering general currency at around this time. Aiken made an important contribution to the proliferation of the word in this sense, but he does not seem to have originated it.

Aiken specified that the actual data were to be inputted on punched cards, with the result of the calculations either being recorded on such cards or by an electric printing mechanism not dissimilar to the printing mechanism of the automatic tabulators. The actual calculations were to be carried out by electromechanical relays; that is, mechanical switches operated by electrically activated magnets.

With this formidable array of hardware and software, the machine could be instructed to carry out any number of complex calculations by means of its paper tape program being left to run until the calculation was completed. Once it was set in operation, the machine would often run for hours or even days, doing what Aiken liked to describe, with a sort of affected rusticity, as 'makin' numbers'.

Many programs were repetitive in their basic nature—computing the successive values of a mathematical table was a typical example. A repetitive program was usually inputted into the machine in the form of a loop of paper tape with the ends glued together. A short program might contain a hundred or so instructions along the paper tape, with the whole loop repeating about once every minute. A longer program would take correspondingly longer to run, and might involve a loop rotation of many hours. Because a long program involved a very lengthy loop of tape, special racks and pulleys had to be built into the machine to take up the slack of the loops of tape and to ensure that the tape was at the right level of tension when it entered the tape-reading unit. The machine might take as long as half a day to calculate one or two pages of a volume of mathematical tables. This length of time seems ludicrous to us today, but at the time represented an enormous advance from manual calculation machines.

Above all, Aiken's computer was remarkably reliable; there was no feeling generated among its users that its results needed to be continually checked. The machine consequently at last did away with the psychological insecurity that had infested the calculation of mathematical tables since Babbage's day.

By the time Aiken's computer was reaching completion, it was becoming known as the Harvard Mark 1. Aiken favoured this name, although IBM preferred the appellation of Automatic Sequence Controlled Calculator. This minor disagreement over the name indicated a wider and growing lack of agreement between Aiken and IBM as the world's first true computer started to become a reality. Aiken was arguably more at fault here than IBM; he was inclined to see the machine as entirely his own invention, while paying insufficient homage to IBM's enormous investment of time and money in the actual building of it. This eventually led to serious problems in his relationship with Thomas Watson.

In many respects the computer which Aiken designed and IBM built did indeed represent the fulfilment of Babbage's dreams, given that the fulfilment used a different technology to the purely mechanical cogwheels Babbage had envisaged. In fact, in one important respect Aiken's machine fell seriously *short* of Babbage's plans. Aiken's computer was incapable of carrying out what is now referred to as a 'conditional branch'—that is, a change in the progress of a program according to the results of a previous computation. The Harvard Mark 1 could not change the direction of a computation in this way, which made complex programs very long and slowed down the machine significantly. This was in spite of the fact that the technology would certainly have made a conditional branch system possible. As Martin Campbell-Kelly and William Aspray point out in their book on the history of the modern computer:

> If Aiken had studied Babbage's—and especially Ada Lovelace's—writings more closely, he would have discovered that Babbage had already arrived at the concept of a conditional branch. In this respect, the Harvard Mark I was much less

impressive than Babbage's Analytical Engine, designed a century earlier.

But the Harvard Mark 1 *had* been completed to fully working order, and the Analytical Engine had not.

On 7 August 1944, by which time there was no longer any real doubt that the Allies would win the war, the press office of Harvard University issued the following announcement:

World's Greatest Mathematical Calculator

The world's greatest mathematical calculating machine, a revolutionary new electrical device of major importance to the war effort, will be presented today to Harvard University by the International Business Machines Corporation to be used by the Navy for the duration.

The apparatus will explore vast fields in pure mathematics and in all sciences previously barred by excessively intricate and time-consuming calculations, for it will automatically, rapidly and accurately produce the answer to innumerable problems that have defied calculation.

The machine is completely new in principle, unlike any calculator previously built. An algebraic super-brain employing a unique automatic sequence control, it will solve practically any known problem in applied mathematics. When a problem is presented to the sequence control in coded tape form it will carry out solutions accurate to twenty-three significant figures, consulting logarithmic and other functional tables, lying in the machine or coded on tapes. Its powers are not strictly limited since its use will suggest further developments of the mechanism incorporated.

The press release also made much of the physical dimensions of the machine and its enormous number of components. As the announcement proudly explained:

The machine is of light-weight, trim appearance: a steel frame, fifty-one feet long and eight feet high ... bearing an interlocking panel of small gears, counters, switches and control circuits. There are 500 miles of wire, 3 000 000 wire connections, 3500 multiple relays with 35 000 contacts, 2225 counters, 1484 ten-pole switches and tiers of 72 adding machines.

The formal presentation of the Automatic Sequence Controlled Calculator to Harvard University, which lent it to the US Navy for the remainder of the War, took place in the Faculty Room of Harvard's University Hall on Monday 7 August 1944. It was a major event. The Governor of Massachusetts was there, as were four admirals, numerous other officers of the armed forces, faculty members, deans, and members of the Harvard governing boards, plus three of the IBM engineers who had constructed the Harvard Mark 1 and members of the machine's operating staff. Aiken, the hero of the hour, made a speech that not only summed up much of his life's work but can also be seen as a summing-up of much of the story of Jacquard's Web.

I should say that the purpose for which we are forgathered here this afternoon is as old as civilisation itself. Our purpose is to consider a device designed to assist in the solution of mathematical problems, and to derive numerical results, because the record is clear that those who invented the fundamental processes of arithmetic were themselves the first to feel the need of mechanical aids.

The first mechanical aid to be used was, of course, the ten fingers of the hands. It is for that reason that the number system we still use at the present time is based on the numeral 10. After the invention of zero, and the extension of the numbers system, the fingers no longer sufficed for counting, and pebbles were used assembled in piles on sheets for the purpose. It was from this that the invention of the abacus came, wherein beads strung on wires took the place of the pebbles, and the wires facilitated their easy movement.

Aiken went on to provide a fulsome and expansive tribute to Babbage. He explained how Babbage attempted

> to build first what he termed a Difference Engine, and then what he called an Analytical Engine, after he had failed with his Difference Engine in the first place. I say Babbage failed, but I should like to make it especially clear that he failed because he lacked machine tools and electric circuits and metal alloys, but through no fault of his own. Babbage's failure was due solely to one fact: he was a hundred years ahead of his time.

Aiken's remarks about Babbage—the father of the modern computer talking about the achievement of someone who might reasonably be described as the father of the entire concept of computing—set the scene for the way Babbage was seen by posterity during the second half of the twentieth century. Of course, blaming Babbage's failure on a lack of 1940s technology rather tends to exalt that technology, which may have been the idea. In fact, as we have seen, there is compelling evidence that under the right circumstances Charles Babbage might indeed have been able to build a working Difference Engine in his own century. The idea that he failed simply because he was ahead of his time is really an oversimplification, though not an unreasonable one.

Aiken continued:

> But after machine tools had been developed, and all the assets of modern manufacturing became theirs, the problem again was opened up by a variety of different manufacturers. Still, the problem remained in the field of the four processes, the four fundamental processes of arithmetic. And then there came one of the fundamental inventions of all times in the art of computation, the use of punched cards for the storing of numbers and for the rapid distribution of those numbers into counters for carrying on numerical processes. It was this

invention, developed by the International Business Machines Corporation, and all the associated parts of mechanisms for speeding and using such cards, that has brought the possibilities of scientific calculating machinery again into a position where one could look at the situation with hopes of success.

This was the situation then, as we saw it, compared with the situation that Babbage saw one hundred and more years ago.

Aiken unfortunately skates over the enormously important contribution of Herman Hollerith as a pioneer in punched-card information storage and also as a key link between IBM and the Jacquard loom. Indeed, Jacquard was also missing from this concise account of the history of mechanical aids to calculation in general, and the history of the punched card in particular. But Aiken had mentioned Jacquard by name in the proposal which he prepared for IBM, so there is no doubt that he was aware of the importance of Jacquard's work as a precursor to Babbage's use of punched cards. In fact, as Aiken was so heavily influenced by Babbage, and as Babbage freely borrowed Jacquard's ideas, it is entirely reasonable to say that Aiken was in a very real sense influenced by Jacquard.

In 1946 Aiken was co-author of an important work entitled *A Manual of Operation for the Automatic Sequence Controlled Calculator*. It is clear to any reader of the *Manual of Operation* that Aiken considered Babbage to be his intellectual father. The very first chapter starts with the quotation from Babbage's *Passages from the Life of a Philosopher* that heads this chapter, a quotation in which Babbage speaks with uncanny foresight to whoever will be the inheritor of his mantle.

Aiken proceeds to look at the most ambitious aspect of Babbage's work:

Having been unable to complete the difference engine, Babbage embarked upon the creation of a far more ambitious concept, an 'analytical engine'. Though the terms of the problem proposed were enough to stagger the contemporary imagination, he attempted to design a machine capable of carrying out not just a single arithmetical operation, but a whole series of such operations without the intervention of an operator. The numbers in the first part of the machine, called the 'store', were to be operated upon by the second part of the machine, called the 'mill'. A succession of selected operations was to be executed mechanically at the command of a 'sequence mechanism' (a term unknown to Babbage). For this latter, he intended to use a variation of the Jacquard cards.

These cards, the precursors of Hollerith's punched cards, were used by the Jacquard weavers to control the looms to produce and reproduce the patterns designed by the artists. The designs were first sketched as they were to appear in the finished product, transferred to squared paper and used as guides for punching cards. The cards allowed certain needles to be extended through the punched holes, thereby controlling hooks which, in turn, raised particular warp threads to produce the desired pattern. In order to continue the weaving of the same design, the cards were interlaced with twine in an endless sequence so that one card was brought into position immediately after another was used. Holes were punched for the lacings as well as for the pegs which guided the cards over a cylinder.

In adapting these cards for use in his machine, Babbage required two decks: one of variable cards and one of operational cards. The first set was designed to select the particular numbers to be operated upon from the store; the second set, to select the operation to be performed by the mill. The deck of operation cards therefore represented the solution of a mathematical situation independent of the values of the parameters and variables involved. Thus the analytical engine was to have been completely general as regards algebraic operations.

Aiken's account of Babbage's work is concise and fairly accurate, even in the light of the very considerable new information about it which has since come to light.

A review of the *Manual of Operation* that appeared in the British scientific journal *Nature* was published under the title 'Babbage's Dream Comes True'. The start of the enormous surge of renewed interest in Babbage's work—an interest that is continuing to increase—dates from around this time. The idea that a brilliant and unjustly neglected Victorian gentleman had anticipated the development of the modern computer by more than a century, and even provided a blueprint for its development, was too romantic, exciting, and interesting to be overlooked. Yet one must avoid being too carried away by it. As I. Bernard Cohen, Aiken's biographer and a leading computer historian, wrote in 1999:

> In evaluating Babbage's possible influence on Aiken, one must keep in mind the enormous gulf between today's image of Babbage as a computer pioneer and his relative obscurity before the 1940s. Today every historian of technology, every computer scientist, and every computer buff is aware of the prescient nature of Babbage's ideas and of the machines he either proposed or constructed in full or in part ... But in the 1930s Babbage was a rather obscure figure whose ideas were generally unknown to working engineers and applied mathematicians. His writings had been out of print for many decades and were to be found only in large research libraries.

Today, the Harvard Mark 1 is no longer entirely intact, although a large, non-working part of its mechanism is in the Smithsonian Institution in Washington DC. The importance of the Harvard Mark 1 as the world's first automatic computer can hardly be overstated, but it must also be conceded that in many respects it was a technological dead end much as Babbage's Analytical Engine was. Aiken's computer was used in a number of applications for the United States Navy, mainly based around calculating mathematical tables. Aiken himself had great hopes for the importance of his computer in the creation of mathematical tables. Eventually the Harvard Mark I was used to produce twenty-five volumes of these tables. But they had a limited use and a relatively short shelf life.

The reason for this was that even during the period when the Harvard Mark I was being built, a new and much better way of building computers was being developed. It involved the use of purely electronic components with no moving parts: a striking contrast to the cumbersome and slow electromagnetic relays. The development of electronic computers is the last episode in the story of Jacquard's Web, or at least so far. These electronic computers soon became so powerful and rapid in their operation that compiling mathematical tables ceased to be important because whenever a particular value was needed, it was much easier to instruct the much faster electronic computer to provide the value in a fraction of a second.

Nor did the Harvard Mark I ever meet Aiken's hopes for a machine that could perform all the calculations required by advanced science. The machine was reliable and worked perfectly well, but during the building process there seems to have been an acceptance on the part of its engineers that it was never going to be as fast in its operation as Aiken had hoped. It was true that—as Aiken liked to claim—the machine was more than a hundred times faster than a human 'computer' and that it could perform six man-months of computing in a day. However, within a few years electronic computers were making this level of performance look slower than the growth of a cactus.

Furthermore, Aiken unfortunately made a serious error of judgement at the dedication ceremony for his computer. He neglected to acknowledge the enormous importance of IBM's role in making the machine happen. It appears that he was so carried away with his own sense of historical destiny that he had forgotten about the reality of the struggle to build the machine. Thomas Watson was incensed by Aiken's lack of gratitude. A generous man by nature, Watson expected generosity and loyalty from those whom he had assisted, and at this crucial moment in Watson's life and in the history of technology, Aiken let him down. Watson called it betrayal. He was infuriated, but he was a man whose emotions quickly tended to coagulate into some course of practical action. Aiken's short-sighted approach may actually have been a disguised blessing for the computer industry, for it led Watson to seek revenge over Aiken by ensuring that IBM soon recaptured the spotlight by building something even better than the Harvard Mark 1.

The better machines that IBM and other organizations created carried the computer forward into realms unimaginable even by a genius such as Charles Babbage. Aiken, who based his machine on Jacquard's and Hollerith's technology and Babbage's inspiration, was about to be superseded. Machines of almost unimaginable power would soon be built that would change the world for ever. And yet the punched card, so far from being abandoned as part of an obsolete technology, was about to enter its heyday.

·16·

Weaving at the speed of light

I'll put a girdle round about the earth
In forty minutes.

Puck in *A Midsummer Night's Dream*

The history of a successful invention is like the story of a river as
it flows from its source to the ocean.

The invention starts out at the source as an idea born in the
mind of a remarkable innovator whom we may be entitled to
regard as a genius. Gradually, like tributaries joining the main
flow of the stream, other inventors and thinkers become
involved. Finally, as the river approaches the ocean, it widens into
a delta. There, hundreds or even thousands of expert technicians
toil away in well-organized, well-funded, yet comparatively
anonymous groups such as the research and development teams
of major international corporations. The river's flow into the
ocean represents the idea's comprehensive acceptance by the
mainstream of human culture and society.

Along the way, a number of possible new tributaries are very
likely to show promise for a while before drying up. Charles

Babbage's plan of building a computer using cogwheel technology was brilliant from a conceptual point of view, and posterity has benefited from his insights—enhanced by Ada Lovelace—into the potential of the computer. Purely mechanical cogwheels, however, proved to be an inadequate technology for enabling a computer to attain the potential which Babbage instinctively knew was possible.

Similarly, the electromagnetic technology of Howard Aiken's Harvard Mark 1 was also consigned to technological oblivion. In practice, the mechanical element of electric relays—switches that physically move from the 'off' to the 'on' position when triggered by an electric current—limited how fast the relays could operate. Like any other mechanical device, they were prone to jamming and overheating. They were also comparatively large, meaning that any machine comprising many thousands of relays had to be enormous. Aiken's computer filled a very large room, and its processing speed was only modest. More ambitious machines could only get bigger. By 1944, when the Harvard Mark 1 was launched to the world, the technology on which it was founded was already starting to seem inadequate to those who understood just how slowly electromagnetic relays operated, and how prone they were to faults and breakdowns.

Was there a way to improve the processing time of computers and reduce the sheer size of the equipment? It seemed at least possible there might be.

In 1881, a young engineer, William J. Hammer, who was working for the great inventor Thomas Edison at Edison's laboratory in Menlo Park, New Jersey, made an accidental discovery that turned out to be of great importance.

At the time, Edison was at the forefront of the struggle to produce a reliable and commercial electric light bulb. The essence of his idea was that a filament would glow within a glass sphere from which the air had been excluded. The exclusion of air was essential to maximize the life of the filament. The work of Edison's team focused on finding the best material for the filament, and the most efficient way to exclude the air. Edison's employee Hammer made the curious discovery that under certain conditions of vacuum and voltage a bluish glow appeared in an evacuated light bulb, suggesting the existence of an unexplained current between the bulb's two filaments in a direction *opposite* to that of the main current.

Despite his inability to understand the phenomenon, initially known as 'Hammer's phantom shadow', Edison was never one to miss an opportunity. He immediately took out a patent for the discovery, prudently renaming it the 'Edison Effect'. But Edison's time was soon fully absorbed in other projects, and as the Edison Effect seemed to have no immediate commercial applications he chose not to investigate it any further.

Other scientists, however, heard about the Edison Effect and, profoundly intrigued by it, started to try to understand what was going on inside the vacuum tube. In due course, this new research concluded that the Edison Effect was due to the surprising fact that there was a passage of electrons—the particles from which electricity is considered to be made—from the negative terminal (the cathode) of the filament inside the tube to the positive terminal (the anode). A tube in which this process was taking place became known as a 'thermionic tube', or more colloquially a 'vacuum tube' or 'valve'. It looked very much like a light bulb, but was designed to maximize the strength of the electron flow rather than actually to produce light.

The discovery of the Edison Effect was to prove one of the crucial discoveries that has made the modern world possible.

The origin of the Effect in the quest for a reliable electric light bulb is yet another example of how one area of technological exploration may 'accidentally' create a gateway to another area, and may even create a completely new technology. The link between electric light and the science of electronics is as momentous as the one between weaving and computing.

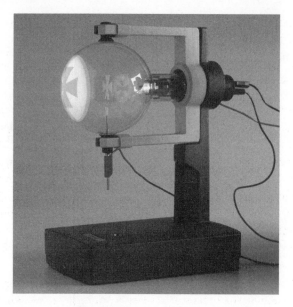

The 'Maltese Cross' experiment.

In time, some scientists wondered whether a thermionic tube could be developed as a new kind of switch. The illustration above shows the 'Maltese Cross' experiment, which projects a beam of electrons within a vacuum tube against a screen at the far end of the tube. The Maltese cross object placed in the path of the cathode rays makes them cast a shadow on the screen. This experiment proved that electrons were particles with a negative charge and a finite mass. These features of electrons meant they could be used in switching devices.

Innovation and refinement of the vacuum tube continued throughout the 1930s and 1940s. When World War II broke out, the possibility of using vacuum tubes as rapid switching devices in computers that would handle calculations with unprecedented speed was vigorously explored around the world—especially in the United States and Europe. Experts hoped to employ electronic computers, among other things, for the rapid decryption of coded enemy communications traffic and the solution of complex calculations relating to the velocities of projectiles. Not for the first time in history, war was proving a powerful catalyst for technological evolution.

Howard Aiken had stolen a march on his rivals in the new field of computing, but only just. The implications of the vacuum tube as a switching device turned out to be extremely important for the history of the computer. Even while the Harvard Mark 1 was in the final stages of being completed, two brilliant engineers from the University of Pennsylvania—J. Presper Eckert and John W. Mauchly (another two intriguing names)—were engaged in building a machine they christened the Electronic Numerical Integrator and Calculator (ENIAC).

Eckert had first showed an interest in electronics when he started building working radios at the tender age of five. After earning a degree in electrical engineering from the university of Pennsylvania in 1941, he was offered a graduate fellowship at the Moore School of Electrical Engineering at the same university. There, he and his professor, Mauchly, made numerous valuable contributions to improving the electronic equipment then available.

The quality of their work became known to the United States Government. They were awarded a contract to construct a digital computer from electronic components. During the construction phase they decided to name this ENIAC. It possessed the enormous technical advantage of not using any electromagnetic relays, but only vacuum tubes. Punched cards were still used to program ENIAC, but its processing speed was much faster than any of its predecessors.

This computer was designed for work analysing ballistics trajectories and producing new tables that allowed a gunner to aim with great accuracy, assuming that the gunner knew a range of key parameters—such as weight of projectile, distance of the target, angle of elevation, and so on. ENIAC is justifiably regarded as the world's first all-purpose, all-electronic digital computer, operated by electronic impulses moving along conductive surfaces, or through vacuums, at the speed of light.

In 1948 Eckert and Mauchly went on to establish a computer manufacturing firm. A year later, they introduced the Binary Automatic Computer (BINAC) to the world. This stored information on magnetic tape rather than on punched cards. Their third computer, the Universal Automatic Computer (UNIVAC 1), was even more sophisticated than the BINAC. Its reliability and sophistication won it many commercial customers. Some consider that the development of the UNIVAC started the global computer boom. Between 1948 and 1966 Eckert received 85 patents, mostly for electronic inventions.

The flow of the river produced other significant developments along the way. These all helped to contribute to the evolution of the digital computer, with electronic components eventually being seen as so superior to mechanical ones that eventually the idea of building computers from mechanical components was discarded altogether. Particularly significant developments in the history of computing during the extremely important years between the late 1930s and the end of the war in 1945 were as follows:

1937—the British mathematician Alan Turing published a paper entitled 'On Computable Numbers, with an Application to the *Entscheidungsproblem* [Decision Problem]'. Turing was interested in investigating whether certain mathematical propositions could be shown to be definitively incapable of proof. His investigation consisted of postulating the idea of a special 'universal mathematical machine' that would be able to assess any proposition and make all the calculations necessary to decide whether the proposition was provable or not. His argument tended inescapably to the conclusion that mathematics will always contain some propositions that cannot be proven. His concept of a special universal mathematical machine was, however, regarded as more significant than the complex puzzle whose solution it was designed to facilitate. The machine was purely theoretical, but in essence Turing had laid the foundations for modern computer science. The machine he proposed and specified very precisely in the paper had all the features of a modern computer: a finite program, a large data-storage capability, and a step-by-step mode of mathematical operation.

The 'Turing Machine', as it came to be called, is even now frequently used as a point of reference in basic discussions of automata theory. It also provided an inspiration for the next generation of digital computers that came into being in the 1940s.

1938—Konrad Zuse, a Berlin-based scientist, completed a prototype for a mechanical binary programmable calculator, that is, one which represented numbers in binary code. This enabled any number to be represented as a sequence of 0s and 1s, or as on/off switches.

1939—On January 1 Hewlett-Packard (HP) was founded by William Hewlett and David Packard in a garage. Unsure whether to call their creation 'Hewlett-Packard' or 'Packard-Hewlett', the two founders decided the matter by the toss of a coin. Their corporation eventually became one of IBM's main rivals and remains one of the world's largest designers and manufacturers of computers and other high-technology equipment. The garage is now an HP museum.

1939—John V. Atanasoff of Iowa State College (now Iowa State University) and his graduate student Clifford Berry completed a prototype 16-bit adding machine. This was able to handle a calculation whose result involves any number up to $2^{16}-1$ or 65 535. This was the first machine which calculated using vacuum tubes.

1939—The outbreak of the war spurred many improvements in technology and accelerated the impetus to make better calculation machines and devices that now started to be described more often as computers.

1941—Atanasoff and Berry completed a special-purpose computer designed to solve simultaneous linear equations. This became known as the 'Atanasoff–Berry Computer' ('ABC'). It had sixty 50-bit memory units in the form of capacitors. Its secondary memory was based around punched cards, except that for speed of card production the holes were not punched into the cards but burned into them.

1943—Computers built between this year and 1959 were often regarded as 'first generation'. They were generally based on electronic valves used in conjunction with electric circuits, and with punched cards playing a key role in allowing the devices to be programmed and in facilitating memory storage.

1943—Encryption experts, including the British mathematician Alan Turing, based at the secret Government Code and Cypher School ('Station X') at Bletchley Park, England, completed a device they affectionately named the 'Heath Robinson' after the British cartoonist famous for his drawings of ludicrously complex mechanisms for carrying out simple tasks. This was a special-purpose computer designed solely to break codes. It was essentially a logic-based processing device and worked using a combination of electronics and electromechanical relays. Apart from its importance in facilitating the cracking of enemy codes, it was also of importance as a forerunner of the 'Colossus' computer.

1943—December. The earliest truly programmable electronic computer was first demonstrated in Britain. It contained 2400 vacuum tubes and was christened the 'Colossus'. It was built by Dr Thomas Flowers of the Post Office Research Laboratories in

London to crack the German 'Lorenz' (SZ42) cypher used by the 'Enigma' machines. Colossus, deployed at Bletchley Park, was the immediate successor to the Heath Robinson. It was able to decipher 5000 characters per second. While not a general-purpose computer, the Colossus represented an enormously important advance in computing science. Ten Colossus machines were eventually built, but by the end of the war had been destroyed by a short-sighted British Government which was afraid that the sophisticated technology they embodied might find its way into Soviet hands.

The introduction of electronic computers boosted processing speeds beyond the wildest dreams of even the most visionary of computer pioneers. ENIAC was able to perform 5000 operations per second compared to the three per second of the Harvard Mark 1. Before long it was routine for computers to perform tens of thousands of operations per second.

The global computer revolution was already well advanced when it received another huge boost: the widespread intro-duction of the transistor. The transistor was actually invented in 1947, but more than ten years of development work were needed to make it a viable alternative to the vacuum tube. When the transistor became commercially available in 1959, it trig-gered another vast step forward for computer technology. The transistor made use of the properties of special materials, known as semiconductors, to create electronic switches that did every-thing a vacuum tube (valve), could do, but which used extremely small components that had the advantage of being solid and not requiring the creation of a vacuum. The transistor's much greater efficiency and reliability, far lower power consumption, and much smaller size than the vacuum tube made it greatly superior to the tube and rendered valves largely obsolete, though they still have some uses in certain specialised electronic equipment, such as some TV cameras and oscilloscopes. Also, some hi-fi connoisseurs prefer the sound quality of valve amplifiers to that of transistorised ones.

By using transistors and by taking advantage of important innovations in how memory capacity was built into computer hardware, computer manufacturers were able to produce more efficient, smaller, and faster digital systems. Some of these machines could process up to 100 000 instructions per second.

Yet even this speed appears snail-like compared with the computer processors of today. These processors, known formally as *microprocessors* or informally as 'microchips' or even 'chips', are basically fantastically miniaturized assemblies of tiny transistors. The Intel Pentium chip, for example, started out containing close to one billion transistors and successive new models of Pentium have even exceeded this. Such chips are far too small to be built manually; they are in fact constructed in completely clean, dust-free environments using powerful microscopes, with the chip itself being etched out of silicon (hence the name 'silicon chip') by powerful rays of light. These chips allow computers to operate at prodigious speeds that show no sign of decreasing or even flattening out.

Meanwhile, punched cards became the indispensable programming medium for almost every computer in the world. Cards were cheap and convenient, and as long as they worked there was no reason to look for another solution. The cards of the 1940s and 1950s were thinner than before. As computers became more sensitive, cards were manufactured and punched with phenomenal accuracy. However, they were still recognizably the direct descendants of Jacquard's cardboard cards for 'programming' a loom.

As processing speeds became faster and faster, experts feared that it would soon not be possible to use punched cards for programming electronic computers. Processing speeds were becoming so fast that even the fastest punched-card feed system, in which punched cards raced far too fast for the eye to see, could never have loaded the cards into the computer's memory rapidly enough to keep pace. Since there was no alternative way to program the electronic computer, the problem actually delayed the evolution of new technology in the late 1940s. During this time, programs had to be 'loaded' into computers by an operator who physically made changes in the wiring of the machines: a tedious, slow, and laborious task that significantly reduced the advantages offered by the new technology.

Fortunately for the computing industry, the problem was solved by the development of a program-reading technique known as *stored programming*. This allowed the computer's memory to hold both the data *and* the program: that is, the raw information and the instructions for processing it. With stored programming, it did not matter that a punched-card system could never keep pace with the processing speed of an electronic computer. The program could be loaded in the form of punched cards, and stored throughout all the processing that followed. Today, stored programming is something we take for granted. All the programs and applications needed can be kept 'in' the computer on a permanent basis. It is now difficult to imagine a time when a computer had to read a pile of punched cards every time it was used.

The development of stored programming, and the tens of thousands of other computing breakthroughs that have made the computer what it is in our world today, were carried out by teams of computer engineers working for large computing corporations. Ever since the late 1950s, this has tended to be the pattern for breakthroughs in computing: they have been the result of collaborative and joint effort by large teams composed of often anonymous people rather than by individual pioneers. This is the river approaching the ocean.

In the 1960s, a new generation of punched cards was born. These cards featured small, usually rectangular perforations. They were read electronically when the perforations either admitted or blocked impulses of light that triggered light-sensitive cells. The figure overleaf shows what a typical punched card from this period looked like. The cards generally continued to incorporate Hollerith's 'missing corner' feature, guarding against them being inserted the wrong way round. Typically, programs would consist of many hundreds or even thousands of punched cards, each one containing one line of the complete program. Many computer users now in their fifties and sixties have nostalgic memories of loading punched cards into computers each time the program was run.

The principle was essentially the same as for the Jacquard loom, where one punched card was needed for each pick of the warp thread. Punched cards continued to be the main medium for loading programs into computers and for inputting data until the mid-1970s, when they were gradually replaced by magnetic tape and magnetic (or 'floppy') disks. Yet it was only in the mid-1980s that punched cards started to become obsolete in the computer industry. IBM manufactured the last one in 1984, a date that is surprisingly recent, considering how powerful and advanced computers had become by then.

Punched cards may now be obsolete technology in computing, but they have not disappeared from other application

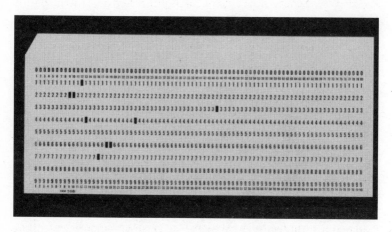

A punched card from the 1960s. The small rectangles are the punched holes.

areas. Even today they are still used for programming older machine tools, particularly printing presses. Their importance in the operation of mechanical voting machines—a curious echo of Herman Hollerith's use of them in census-taking—continues, especially in the United States. The voter punches out a hole on a printed card to indicate his or her chosen candidate. The card is then either handed to an attendant for manual counting or placed into a vote-counting machine, which reads the card mechanically or using a light pulse. In many US states, the sheer expense of upgrading to optical scanners or touch-screen vote technology means that punched-card voting machines are likely to be used for many years to come.

This voting system received huge international attention late in the year 2000 during the neck-and-neck US Presidential Election. Imperfectly and mispunched cards suspected of possibly providing false results in Florida and conceivably also in some other states were debated by the news media for days. In retrospect, the problem was not that the system failed the US electorate, but that the presidential race was almost unbelievably tight, making every spoiled vote a cause for major debate. As Richard K. Scher, a professor of politics at the University of Florida, said in an e-mail he sent to me:

> While I fear the media very much overplayed the difficulty of interpreting some of the cards, when the difference between the candidates was only 537 votes out of nearly six million, the difficulties in reading the ambiguous ballots became central.

Whatever the eventual fate of punched cards as a way to record and store information, the immortality of the idea behind Joseph-Marie Jacquard's loom cannot be questioned. He created a brillant system, enabling the action of a complex mechanism to be changed infinitely according to an endlessly variable set of instructions. Indeed, as we have seen, the entire concept of 'hardware' and 'software' can be traced directly to Jacquard himself. Punched cards are no longer part of the infrastructure of the computer industry, but their influence lives on in the very way in which we make computers work for us.

There are even some technical echoes of the historical importance of punched cards in the most sophisticated modern computers. For example, if you open an MS-DOS prompt window on a personal computer the window is 80 columns wide. This comes from the standard Visual Display Unit (VDU) size of 80 columns in width, which in turn was originally designed to allow the display of a complete 80-column punched card. Another example is the Extended Binary Coded Decimal Interchange Code (EBCDIC) character set still used on mainframes and on some other computers. This character set is heavily influenced by punched cards; there are 'gaps' between the letters I and J and between R and S, just as there were on the now superseded punched cards.

These technical links emphasize the intimacy of the relationship between punched cards—originally designed for weaving—and state-of-the-art computers.

The advance of technology has, literally, allowed us to weave at the speed of light.

·17·
The future

I don't like trying to predict the future. Historians can't predict the future any better than anyone else.

Thomas Bergin, editor-in-chief,
Annals of the History of Computing

The varied and intimate connections between weaving and computing come into a sharper focus, and become increasingly extensive, the more we probe into the relationship. Here is another intriguing link.

You can see it every time you peer at a computer screen. A computer is programmed to store, retrieve, and display images using *pixels*: an abbreviation for the term 'picture element'. A computer screen is divided into tiny discrete units, each of which may or may not be illuminated (whether in monochrome or colour) in order to form part of the image.

The pixels are so small that they cannot normally be seen, but they become visible if we enlarge any item on a computer screen, including a word. For example:

Jacquard

Here the letters have been enlarged to a size where their pixel-based structure is clearly visible. The curves of the letter are not curves at all, but a sequence of steps composed from square pixels.

This pixel-based method of representing an image bears a great resemblance to the way the master-weavers of Lyons wove images from silk. This is because the woven images created in silk fabric by the master-weavers of Lyons on the Jacquard looms still used in Croix Rousse are themselves in fact nothing more or less than digital images.

A digital image in this sense is one in which the picture is represented by a code consisting of only two elements. A digital image is made using a representational system that places the image on a grid, with the tiny squares or rectangles of the grid being either filled with a colour (which may include black or white) or left blank. The 'filling in' is one element and the blank-ness is the other. Those are, by the very nature of weaving and computing, the only two options.

The link between weaving and representations of letters on a computer screen can be seen very clearly by looking at how the weavers of Lyons wove words into their designs. In the Museum of the History of Weaving in Lyons, for example, there is a woven tableau made to celebrate the visit of Napoleon to Lyons on 15 January 1802 (*26 Nivôse An 10* by the short-lived 'rational' Revolutionary calendar). According to the message woven on it, the tableau was produced in the presence of Napoleon himself. It is inscribed with a compliment to him—*il nous a donné la paix* ('he gave us peace'). In effect, it is a woven image of

(*right*) Digital writing from 1802.

İL
NOUS
A
DONNÉ LA
PAIX

Fait en présence du premier
Consul
À LYON LE 26 NIVOSE AN 10

digital writing. The letters are composed of little rectangular units, just like the letters on a computer screen. The woven pixels are, in effect, tiny rectangular sections of interlaced warp and weft of which the image is composed.

This remarkable resemblance between a piece of fabric woven in silk and a computer graphic shows yet again that weaving and computing are closely related expressions of the same human desire: to capture reality within a mechanism where the process is, at a fundamental level, limited to just two possible modes: yes or no, colour or blank, stitch or no stitch. Weaving, like modern computing, is indeed at heart a digital process.

Given the historical link between weaving and computing, it seems almost too great or lucky a coincidence that the largest network of computer connections in the world is called the World Wide Web. The link between Jacquard's idea and the World Wide Web is more metaphorical than literal. All the same, the linguistic links between weaving and the amazing global phenomenon of the Internet are irresistibly strong, and thoroughly fascinating. Sir Tim Berners-Lee, the British computer scientist who invented the World Wide Web, published an account of his invention in his 1999 book *Weaving the Web*. In his book, he sets down how the idea first came to him:

> When I first began tinkering with a software program that eventually gave rise to the idea of the World Wide Web, I named it Enquire, short for *Enquire Within upon Everything*, an old book of Victorian advice I noticed as a child in my parents' house outside London. With its title suggestive of magic, the book served as a portal to a world of information, everything from how to remove clothing stains to tips on investing money. Not a perfect analogy for the Web, but a primitive starting-point.

And he continues:

What that first bit of Enquire code led me to was something much larger, a vision encompassing the decentralised, organic growth of ideas, technology and society. The vision I have for the Web is about anything being potentially connected with anything. It is a vision that provides us with new freedom, and allows us to grow faster than we ever could when we were fettered by the hierarchical classification systems into which we bound ourselves. It leaves the entirety of our previous ways of working as just one tool among many. It leaves our previous fears for the future as one set among many. And it brings the workings of society closer to the workings of our minds.

The proliferation of inexpensive, powerful, and portable computers around the world has created a global *web* of connections, like a giant electronic weaving-loom with untold millions of warp threads and unlimited weft threads. Furthermore, the notion of a web with numerous connections is indeed a convenient way of understanding how the story of this book has been told. And what exactly *is* Jacquard's Web? At one level, the web of these links between the Jacquard loom and the modern computer. Yet these links are so strong, and so compelling, and lead so irresistibly to the Internet itself, that it is not stretching credibility too far to describe the Internet itself as Jacquard's Web.

Our story has taken us from the silk industry in Lyons in the late eighteenth century, through to the Industrial Revolution, and on to the enormous expansion of industry in the United States at the end of the nineteenth century. It has covered America's surge to economic supremacy during the twentieth century, the advances in computing during the Second World War and the technological wizardry of the modern revolution in information technology. There is, surely, something both fitting and appropriate that the story ends—at least for the time being—with a global network of billions of interconnected computers known as the World Wide Web. Or, if you prefer, Jacquard's Web.

And so we move into the future.

There has been extensive anecdotal evidence over the past few decades that trying to forecast where computing is going in the future is a close to impossible task even for those who ought to know. Many prominent computer professionals have got it embarrassingly wrong. To take just two examples, Thomas Watson himself once said he doubted whether the world would ever need more than a handful of computers, while the legendary Bill Gates of Microsoft once ludicrously underestimated the maximum amount of memory computers would ever need to contain. Gates even initially failed to spot how important a development the Internet was likely to be. He, and other gurus of the high-tech revolution, have soon come to regret their attempts to peer into the crystal ball of the technological future. This, like Harry Potter's Mirror of Erised, has generally turned out to reflect their own current desires rather than to show what is really going to happen in time to come.

A serious difficulty with making reliable predictions about the future of computing is that the technical breakthroughs on which significant developments depend so often happen by accident, or as serendipitous spin-offs from some other process. William Hammer's accidental discovery of electron flow in Thomas Edison's workshop when trying to build a more efficient light-bulb is a typical, and classical, example of this process at work. This, and breakthroughs like it, are not only fundamentally unpredictable but also very often *unimaginable* until they have actually taken place.

Furthermore, even when a scientific or technical breakthrough has facilitated an important initiative in the computer industry, experience shows that the subsequent path of that initiative is often unpredictable too. This is another lesson the story of Jacquard's Web teaches us: that the progress of a new technology, far from following a clear, logical track, is generally a haphazard and even messy affair. Its success or failure itself will depend on a complicated web of dynamic, complex factors such as practical necessity, financial pressures, political considerations, and the personal needs and prejudices of the would-be inventor and of everybody else whose decisions affect the invention's progress.

No account of the history of the computer, and its likely future, would be complete without a discussion of the concept of 'artificial intelligence' (AI). The notion of AI takes the evolution of the Jacquard loom to its logical conclusion. If it is possible to build a special kind of Jacquard loom that can weave information, why not build a special kind of information processing loom that can think entirely for itself? And if this is possible, could not an even more brilliant type of loom be built that could think for itself about *anything*?

Since the 1950s, there has been considerable discussion of the possibilities for the practical realization of AI. The term appears to have been originated by the British computer pioneer Alan Turing. In its most ambitious form, AI means machine intelligence that could, at least in theory, seek to imitate the thought-processes of the human brain.

Naturally enough, the notion of an intelligent computer— and in particular a computer that suddenly becomes self-aware and intelligent—offers many exciting dramatic possibilities, especially if the intelligence turns out to be of the malevolent kind. The idea has inspired countless novels and movies, most notably Arthur C. Clarke's novel *2001: A Space Odyssey* and Stanley Kubrick's 1969 movie of the same title.

One particular sequence in the movie features a spacecraft accommodating a revolutionary computer that exhibits self-awareness and a level of intelligence not only comparable with a human being's, but in excess of it. During the mission, the computer makes an unaccountable mistake that triggers a spiral of further errors and disobedience. The captain 'tells' the computer he will disconnect it if it does not obey it. In response, the computer, dreading disconnection above everything else, sets out to kill the crew and comes close to doing so. The story is effective because the computer's motivation seems entirely plausible. Certainly, if a computer which could think for itself *were* ever built, it would be more than reasonable for us to suppose that one of its first concerns would be to take any action necessary to prevent itself ever from being switched off.

After *2001* was released, it was pointed out that the computer's name, HAL, was simply the acronym IBM, with each letter being substituted for the one that precedes it in the alphabet. Had Arthur C. Clarke deliberately intended this to be the case? To try to clear the matter up, I contacted Clarke while researching this book. He e-mailed me by return to emphasize that the abbreviation HAL was simply taken from 'Heuristic Algorithmic': the technical term for the basis of HAL's programming. The 'connection' with IBM is thus apparently a coincidence, albeit an intriguing and convenient one: perhaps even possibly an unconscious one on the part of the author. This ghostly link between the name of the world's first—if fictional—intelligent computer and the global corporation which, via the Hollerith link, can reasonably be said to owe its origins to the idea behind Jacquard's loom, resonates in the history of computing like the fulfilment of a prophecy.

Fictional representations of artificial intelligence are so effective and thought-provoking that the fact of the minimal real-life progress made in the field can seem, by comparison, acutely disappointing. The simple fact is that today, no computer on Earth offers a level of performance that *remotely* resembles the reasoning and creative power of the human mind. Even the problem of mimicking the human brain's remarkable skill at visual recognition—a task that initially seemed to offer the prospect of being achievable—has proved much more difficult than was originally anticipated.

Why can we imagine intelligent machines but not make them? Part of the answer appears to be that our technological imaginations always seem to run well ahead of our technological capabilities. This seems to be a consequence of the nature of our tool-maker's brain. Very likely if it wasn't the case, we would not have achieved a fraction of what we have achieved as tool-making creatures.

At a technical level, a fundamental and overwhelming problem with developing true AI is that it is not known precisely how the human brain 'digitizes' information in order to make it available as the subject of thought carried out by brain cells. There is a certain limited knowledge of the coding used by nerve fibres and even how some of the signal processing (for tasks such as visual processing) is performed, but otherwise our information about how our own brains actually work is woefully incomplete. Until we have a much more comprehensive knowledge of this, it is difficult to imagine how a computer could be taught to understand even basic facts about our world; facts that the machine would need to know if it was to display any useful intelligence.

There has, however, been progress with AI in areas where the field in which the 'intelligence' is to be deployed is of its essence restricted and specialized and consequently does not require the computer to have any knowledge of the natural world. In particular, programs designed to play the game of chess have been notably successful.

A significant milestone was reached in 1996. During a match played in Philadelphia that year between the then World Chess Champion Gary Kasparov—widely regarded as the best chess player of all time—and an IBM-sponsored chess-playing computer program named Deep Blue, Kasparov suffered a loss to the computer. This was the first loss ever by a reigning World Chess Champion to a chess program at normal tournament time limits. Subsequently, in May 1997, an updated Deep Blue with a capacity for considering 200 million positions per second—twice its normal speed—beat Kasparov in a six-game match, with the computer scoring two wins, three draws, and one loss.

Strictly speaking chess-playing programs do not consciously 'think' in any meaningful sense. What they are able to do is to assess tens of millions, or hundreds of millions, of possible chess positions within a fraction of a second. Nonetheless, the program does display a form of artificial intelligence, because with approximately twenty possible moves available, on average, to each player for each move, the number of potential positions in a game of chess after only ten moves by each side is already astronomical. It is totally impossible for any computer to be programmed with details of all the positions that could ever arise on a chessboard. Instead, the programmers have no choice but to equip the computer with a set of rules that it can use in any given position. In other words, the computer is obliged to make its own decisions having been programmed with initial instructions. This is not unlike how the human mind behaves, at least when it is solving problems.

All the same, we are entitled to ask ourselves just how intelligent computer chess programs really are if we consider that such a program, operating inside a burning building, would simply go on playing chess rather than trying to escape. But strictly speaking that would be our fault for not giving the computer any way of knowing that the building was on fire.

In any event, chess-playing computer programs nowadays routinely compete against human chess grandmasters (and against other chess-playing programs) in international tournaments. Many human expert players have developed successful strategies for dealing with chess-playing programs by keeping the position blocked and avoiding the kind of tactical complications at which computers, which don't usually make mistakes in tactical calculations, excel. Whether this kind of mental struggle for supremacy between man and machine intelligence in the limited arena of the chessboard will one day be enacted in the greater world is surely one of the most terrifying questions we face as we move ahead into the twenty-first century and beyond.

Meanwhile, we have little choice but to face the fact that it is extremely risky, if not downright foolhardy, to set down specific predictions for how computers—those modern incarnations of Jacquard's loom—are likely to develop in the future. But we can be certain that the same qualities of ingenuity, passion to make a dream come true, and sheer desire to solve a pressing practical problem that have played such an important part in driving forward the story told in *Jacquard's Web* will fuel developments in the future, developments that are sure to surprise and delight us. The Jacquard loom has never stopped weaving since that first day two hundred years ago. How can we imagine, as we move on into the future, that it ever will?

THE END

Appendix 1

Charles Babbage's vindication

Charles Babbage's belief that the Difference Engine could indeed be built and made to work properly was given a triumphant vindication in 1991. In that year a team of engineers at the Science Museum in South Kensington, London completed an astonishing six-year project to build Babbage's second Difference Engine design: a modified and improved version of the Difference Engine prototype known by Babbage as Difference Engine No. 2. The machine built in 1991 has worked perfectly from the outset.

The Science Museum once offered regular opportunities to view the completed Difference Engine in action. However, at present these have been suspended to allow the Museum's engineers to concentrate their efforts on building the replica of the Difference Engine No. 2 mentioned below. Yet ultimately, the only way of appreciating the true brilliance and the beauty of Babbage's concept of a Difference Engine—and especially the carriage mechanism—is to view the completed Difference Engine No. 2 in action at the Museum.

The one place in the world where you can see a completed Babbage Engine in operation is, at the time of writing, the Science Museum in London, by special arrangement. Only a trained curator can operate the Difference Engine and make the machine perform calculations. Unlike many of the exhibits in the Museum, the Difference Engine cannot be activated by members of the public directly. Turning the Engine's handle requires expert training in order to make the machine operate at the right speed and with the correctly-paced pauses between each turn.

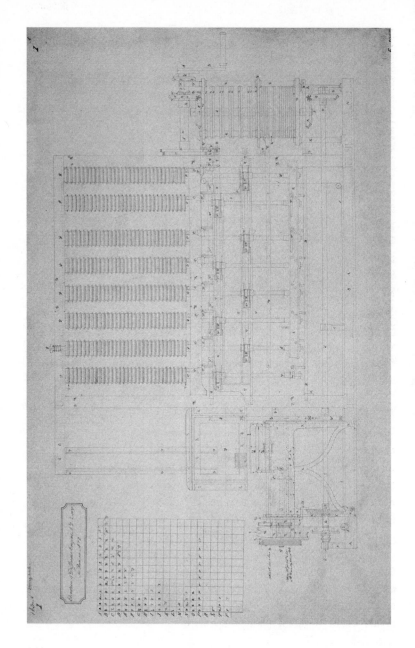

The modern Difference Engine No. 2 completed in 1991 by a team at the London Science Museum to Babbage's plans. It works perfectly.

I have seen the Difference Engine operating on two occasions. Each time the operator has been the Science Museum's then assistant director and head of collections, Doron Swade, the inspired and dedicated computer scientist and historian who led the modern building of Difference Engine No. 2 and who has done so much to make an entirely new generation aware of Babbage's foresight and genius.

When the Difference Engine is operating, the levers of the carriage mechanism—regularly spaced in successive segments of the Figure Wheels one above each other in the columns—create an entirely unexpected phenomenon: a beautiful oscillation that has the appearance of endlessly changing and rippling sine waves flowing around at the back of the machine.

(*left*) Babbage's plans for Difference Engine No. 2.

Watching the machine in action has been among the most soul-shattering experiences of my life. On both occasions I was poignantly aware that Babbage himself never saw his own creation completed, let alone working with complete beauty, precision, and harmony. Witnessing the Difference Engine in operation causes profound questions to leap into one's mind with the regular beat of the cascading cogwheels. Is the machine one sees in operation really a nineteenth-century machine? A twentieth-century creation? A twenty-*first* century device? Or is it something else that exists, both miraculously and uneasily, in a kind of netherworld of the scientific imagination, lurking at the margins of both the nineteenth and twentieth centuries?

Babbage was to enjoy yet one more vindication. In April 2000, another team at the Science Museum, also headed by Doron Swade, completed the construction, to Babbage's specifications, of a printer which Babbage planned to be used in conjunction with the Difference Engine, his first cogwheel calculator. The project to build Babbage's printer was sponsored by Nathan Myhrvold, the former chief technology officer of Microsoft, for whom the Museum is building a replica of Difference Engine No. 2.

The printer automatically produces results of up to thirty digits on a paper roll. The roll can be used for checking the Difference Engine's output. The printer also presses the results in soft metal or *papier mâché* trays from which plates can be made for use in a conventional printing press. The results are produced simultaneously in two print sizes, and there is also a facility for the page layout to be altered. Numbers can be arranged in one, two, or three columns. It is even possible to vary page margins and the spacing between columns. Results can be produced down one column, with an automatic fly-back to the top of the next column, or across the page with an automatic line-rack to the start of the next line. The apparatus is entirely mechanical.

The remarkable success of the team at the London Science Museum in completing the construction of a working Babbage machine, and subsequently a working printer, has naturally generated a surge of fresh interest in Babbage. If any nineteenth-century scientist deserves greater and greater attention from a new generation of anyone interested in how technology has evolved, it is surely him.

Appendix 2

Ada Lovelace's letter to Charles Babbage, 14 August 1843

The following is the full text of the letter which Ada Lovelace wrote to Charles Babbage on Monday, 14 August 1843, offering him comprehensive assistance in connection with his work on the Analytical Engine.

My Dear Babbage. You would have heard from me several days ago, but for the *hot* work that has been going on between me & the printers. This is now happily concluded. I have endeavoured to work up everything to the utmost perfection, *as far as it goes;* & I am now well satisfied on the whole, since I think that *within the sphere of views* I set out with, & in accordance with which the whole contents & arrangement of the Notes are shaped, they are very complete, & even admirable. I could *now* do the thing *far better;* but this would be from setting out upon a wholly different *basis.*

I say you would have heard from me before. Your note (enclosed on Monday with my papers & c), is such as demands a very full reply from me, the writer being so old & so esteemed a friend, & *one whose genius I not only so highly appreciate myself, but wish to see fairly appreciated by others.*

Were it not for this desire (which both Lord L— & myself have more warmly at heart than you are as yet at all aware of), coupled with our long-established regard & intercourse, I should say that *the less notice taken by me of that note—the better;* & it was only worthy to be thrown aside with a smile of contempt. The *tone* of it, it is impossible to misunderstand; & as I am myself always a very "*explicit function of x,*" I shall not pretend to do so; & shall leave to *you* (if you please it) to continue the "i*mplicit*" style which is exceedingly marked in the said note.

As I know you will not be *explicit* enough to state the *real* state of your feelings respecting me at this time, I shall do so for you. You feel, my dear Babbage, that *I* have (tho' in a negative manner) *added* to the list of injuries & of disappointments & mis-comprehensions that you have already experienced in a life by no means smooth or fortunate. You *know* this is your feeling; & that you are deeply hurt about it; & you endeavour to derive a poor & sorry consolation from such sentiments as "Well, she don't *know* or *intend* the injury & mischief if she has done" &c.

You say you did not wish me to "break my engagement, but merely to ask to be released from it." My dear friend, if the engagement was such that I had no right to break it *without leave*, I had still *less right* to appeal to the *courtesy* of parties, in order to obtain an apparent sanction & excuse for doing that which their *justice & sense of their own rights* could not have conceded. There is no greater sin, or deeper *double-dealing* in this world, than that of endeavouring thro' the influence of secondary motives to get the apparent support & consent of others towards that which on a *higher* & more *general* motive would be inadmissible. Your reply to this will be, that my principle is *just, good, & great;* but that it did not apply to the particular case; in as much as the editors would themselves have been glad of an excuse to be released, under the circumstances. You must allow me to state that I took measures for ascertaining that point beyond all possibility of doubt, & that I found it to be very much otherwise.

You will deny & dispute this; or more probably you will immediately perceive grounds *why*, (from various unworthy arrière-pensées) the persons concerned might still wish to retain the article. Remember however, that when you did this, you have *shifted your original ground*; & that your question then becomes not whether Lady L— ought to oblige two parties, *you & the editors*, who both tho' on different grounds wish to dispose of my publication thro' another channel than that originally proposed, but whether Lady L— ought tacitly to lend herself to certain possible or probable unworthy motives entertained by the editors. Now to this the reply is perfectly plain in the opinion of all parties accustomed to fair & honorable dealings, uninfluenced by secondary motives. My engagement was *unconditional*, had no reference to the *motives* of the parties with whom I contracted it. I have therefore no right to withdraw it on grounds *subsequently* thought of. If I had undertaken to do the thing specially *for* you, in addition to its being *for* them, the case would be wholly different. But with the circumstance of your happening to be a private friend of my own, & of my therefore being too happy & delighted to make prior engagement *especially* pleasing & useful to you, *they* had nothing to do. Consequently, because my private friend wished it, (however justly), this could form no *real* equitable ground for withdrawing the article.

I have now touched on all grounds which can be taken on the supposition of its *really being pernicious to your interests* that I have thus allowed the article to appear. This however I cannot agree to or believe; & were you not influenced by a set of feelings which are very different from those that I myself, & the minds whom I most esteem, can consider wise, justifiable, or in harmonious accordance with man's moral nature, *you* would not think so. Mind, I do not say that *your* views may not be in reality higher, juster, & wiser, than my own. But my moral standard, such as it is, I must stick to; as long as it *is* my moral standard. It would not be of any use for me to endeavour making you see thro' *my* glasses; for, (besides the fact that they may be as far or farther than yours, from refracting & reflecting quite truly), no one *can* instanter alter the views & modes of feeling of a *life*. But I *do* wish you to understand the fact, that *I believe* myself (however erroneous that belief may be), to have forwarded *your* interests far more by allowing the article to appear than I should have done by any of the courses you suggested. I *have* a right to expect from you the belief that I do sincerely & honestly take this view. For if *your* knowledge of *me* does not furnish sufficient grounds for doing so, then I can only say that *no* mutual knowledge of any human beings in this life, can give stable & fixed grounds for faith & confidence. Then Adieu to *all* trust, & to everything most generous, in this world!–

I must now come to a practical question respecting the future. *Your* affairs have been, & are, deeply occupying both myself & Lord Lovelace. Our thoughts as well as our conversation have been earnest upon them. And the result is that I have plans for you, which I do not think fit at present to communicate to you; but which I shall either develop, or else throw my energies, my time & pen into the service of some other department of truth & science, according to the reply I receive from you to what I am now going to state. I do beseech you therefore deeply & seriously to ponder over the question how far you can subscribe to my conditions or not. I give to *you* the *first* choice & offer of my services & my intellect. Do not lightly reject them. I say this entirely for *your own* sake, believe me.

My channels for developping [sic] & training my scientific & literary powers, are various, & some of them very attractive. But I wish my old friend to have the *refusal*.

Firstly: I want to know whether if I continue to work *on* & *about* your own great subject, you will undertake to abide wholly by the judgement of myself (or of any persons whom you may *now* please to name as referees, whenever we may differ), on *all practical* matters relating to *whatever can involve relations with any fellow-creature or fellow-creatures.*

Secondly: can you undertake to give your mind *wholly* & *undividedly*, as a primary object that no engagement is to interfere with, to the consideration of all those matters in which I shall at times require your intellectual *assistance* & *supervision*; & can you promise not to *slur* & *hurry* things over; or to mislay, & allow confusion & mistakes to enter into documents, &c?

Thirdly: If I am able to lay before you in the course of a year or two, explicit & honorable propositions for *executing your engine*, (such as are approved by persons whom you may *now* name to be referred to for their approbation), would there be any chance of your allowing myself & such parties to conduct the business for you; your own *undivided* energies being devoted to the execution of the work; & all other matters being arranged for you on terms which your *own* friends should approve?

You will wonder over this last query. But, I strongly advise you not to reject it as chimerical. You do *not* know the grounds I have for believing that such a contingency may come within my power, & I wish to know before I allow my mind to employ its energies any further on the subject, that I shall not be wasting thought & power for no purpose or result.

At the same time, I must place the whole of your relations with me, in a fair & just light. Our motives, & ways of viewing things, are very widely apart; & it may be an anxious question for you to decide how for the advantages & expediency of enlisting a mind of my particular class, in your service, *can* over-balance the annoyance to you of that divergency on perhaps many occasions. My own uncompromising principle is to endeavour to love *truth & God before fame & glory or even just appreciation*; & to believe generously & unwaveringly in the *good* of human nature, (however dormant & latent it may often seem).

Yours is to love truth & God (yes, deeply & constantly); but to love *fame, glory, honours, yet more*. You will deny this; but in all your intercourse with *every* human being (as far as I know & see of it), it is a *practically paramount* sentiment. Mind, I am not *blaming* it. I simply state my belief in the *fact*. The fact may be a very *noble & beautiful* fact. *That* is another question.

Far be it from *me*, to disclaim the influence of *ambition & fame*. No living soul ever was more imbued with it than myself. And my own view of duty is, that it behoves me to place this *great & useful* quality in its *proper relations & subordination*; but I certainly would not deceive myself or others by pretending that it is other than a very important motive & ingredient in my character & nature.

I wish to add my mite towards *expounding & interpreting* the Almighty, & his laws & works, for the most effective use of mankind; and certainly, I should feel it no small *glory* if I were enabled to be one of his most noted prophets (using this word in my own peculiar sense) in this world. And I should undoubtedly prefer being *known* as a benefactor of this description, to *being* equally great in fact, but promulgating truths from obscurity & oblivion.

At the same time, I am not sure that 30 years hence, I may put even so much value as *this*, upon human fame. Every year adds to the unlimited nature of my trust & hope in the Creator, & decreases my value for my relations with mankind *excepting as His minister*; & in *this* point of view those relations become yearly more interesting to me. Thro' my present relations with *man*, I am doubtless to become fit for relations of another order hereafter; perhaps *directly* with the great Power Himself. Of course my view respecting every even *casual* social contact & intercourse, takes a corresponding colour; & will do so increasingly, if *that* view should become more confirmed.

My dear friend, if you knew what *sad & direful* experience I have had, in ways of which you cannot be aware, you would feel that *some* weight *is* due to my feelings about God & man. As it is, you will only smile & say, "poor little thing; she knows nothing of life or wickedness!"

Such as my principles are, & the conditions (founded on

them), on which alone you may command my services, I have now stated them; to just such extent as I think is absolutely necessary for any comfortable understanding & cooperation between us, in a course of a systematized & continued intellectual labour. It is now for *you* to decide. Do not attempt to make out to yourself or me that our principles entirely accord. They do *not,* nor *cannot* at present, (for people's views as I said are not to be altered in a moment).

Will you come *here* for some days on Monday. I hope so. Lord L— is very anxious to see & converse with you; & was vexed that the Rail called him away on Tuesday before he had heard from yourself your own views about the recent affair.

I sadly want your *Calculus of Functions*. So *Pray* get it for me. I cannot understand the *Examples*.

I have ventured inserting to one passage of Note G a small Foot-Note, which I am sure is *quite tenable*. I say in it that the engine is remarkably well adapted to include the *whole Calculus of Finite Differences*, & I allude to the computation of the *Bernoullian Numbers by means of the difference of Nothing*, as a beautiful example for its processes. I hope it *is* correctly the case.

This letter is sadly blotted & corrected. Never mind that however.

I wonder if you will choose to retain the lady-fairy in your service or not. Yours ever most sincerely.

A.A.L.

Appendix 3

How the Jacquard loom worked

The operation of the Jacquard loom can best be understood by looking at a simplified diagram of its operation (see following page). The control device is fixed to the top of the loom. The lifting hooks are indicated in the diagram by a. These hooks pass in a perpendicular fashion through eyes in a number of horizontal needles equivalent to the number of the lifting hooks. The horizontal needles are shown in the diagram as stretching from b to c. They lie in rows in a frame indicated in the diagram by d. In the diagram, in order to simplify the explanation, only eight hooks and eight needles are shown, but in the actual Jacquard loom there are as many as 400 of each, or fifty in each of the eight rows, allowing the machine to control the action of up to 400 warp threads, or even more.

The horizontal needles protrude through the frame at the points indicated by c. They are kept in that position by helical or spiral springs in the position indicated by e. These springs are placed in hollowed-out cavities in the frame. They are held fast inside the frame. Whenever pressure is applied to the tips of the needles at the points c, the needles will slide back into the frame, but the instant the pressure is removed, the elasticity of the springs will make the needles shoot outwards again. The extent to which the needles can protrude from the frame is checked by vertical pins located at g.

In order for the loom to operate, there has to be a system for governing the action of the individual needles that control the lifting hooks which in turn control the action of the loom's warp

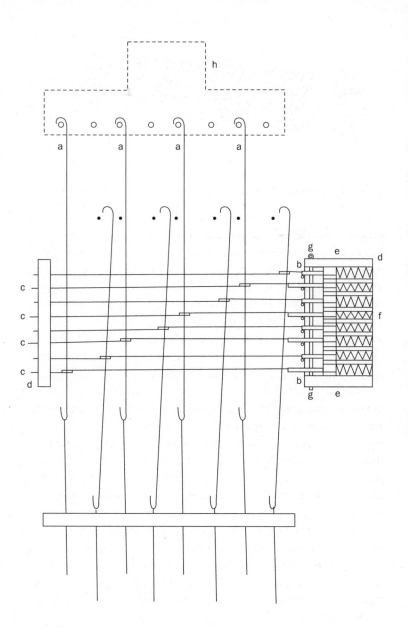

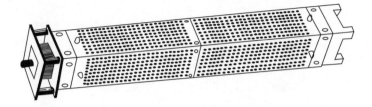

The bar of the Jacquard loom against which the punched cards were pressed.

threads. The device that permits this action to be governed with an enormously high degree of accuracy is a special square bar in the shape of an elongated cube. The bar, shown in the figure above, is fitted with hundreds of tiny identical holes, each of which is the right size to accommodate the tips of the needles. The operation of the bar in the Jacquard attachment is arranged so that the bar turns onto the next flat, perforated side for each pick of the shuttle. Every time the bar turns it momentarily locks against the ends of the horizontal needles.

The mechanism is designed so that as the bar turns onto its next side, one card from the chain of punched cards turns with it and is pressed firmly against the ends of the needles. This pressing action is part of a much more complex overall operation—that is, the entire operation of the Jacquard loom—but each press is in itself complete, comprehensive, and provides a moment when the card is stationary as it pushes against the needles.

The cards themselves are partially perforated. The holes are made in the cards so that whenever a card is pressed against the bank of needles it only causes those needles to be raised which lift the warp thread for a pick of a particular design. Naturally, the number of perforations in each card is less than those on the sides of the revolving bar.

(*left*) How the Jacquard loom works.

Nature depicted by mathematical analysis: the *mise-en-carte* principle in action.

What happens is that when a needle confronts an unperforated area of the card, the needle will not be released and the warp thread it governs will remain down. On the other hand, if a needle confronts a hole, the needle tip will penetrate the hole, ending up in one of the holes on the square bar, and the warp thread the needle governs will be raised.

The perforations in the card are made in order to ensure that the required pattern is woven by the weaver working below. In order for this to happen it is necessary that each card controls the warp threads for each pick of the weft. As a result, where the pattern is large or very varied, the number of cards is correspondingly large. A chain of punched cards in a Jacquard loom frequently reached 2000 in number, and sometimes even exceeded this.

The production of the cards for the composition of different patterns on the Jacquard loom was always a distinct and separate business from the actual weaving. The first step was for the design to be first drawn onto paper on a much bigger scale than that of the finished design. The paper used for this purpose is divided by lines into very small squares: exactly like graph paper.

The Jacquard loom: the ancestor of the computer.

Once the design had been placed onto the ruled paper, the craftsman took a special frame containing more than enough warp threads and cross threads to allow the design to be transferred to it. This often necessitated as many as 500 warp threads and a similar number of cross-threads. The warp threads and cross-threads on the frame amounted to a representation of the ruled paper on a horizontally and vertically threaded frame. A craftsman then proceeded to insert a little piece of thread into all the intersections on the frame that corresponded with the composition of the pattern on the ruled paper.

Each insertion produced just one tiny element of a design which might contain many thousands of such elements. In effect, the craftsmen were constructing a computer graphic on a pixel-by-pixel basis. The French term for this initial mapping work was *mise en carte*: one translation of which is 'put into a card'. That, of course, is a precise description of the essential nature—and extraordinary ingenuity—of the process.

Acknowledgements

Researching and writing *Jacquard's Web* has been possible because many people were prepared to help me, even when doing so made heavy inroads into their own schedules.

I am especially grateful to Tim Bergin, editor-in-chief of *Annals of the History of Computing*. Tim read an earlier draft of this book and made many useful suggestions. I am similarly indebted to Professor Martin Campbell-Kelly, head of computer science at the University of Warwick, for his numerous incisive comments. Inevitably, a book such as this involves taking a considered position on controversial historical issues that remain very much a matter for debate. Neither Tim nor Martin agreed with all of my conclusions, but their suggestions and comments were infused with a spirit of generosity and encouragement which was a delight to receive.

My sincere gratitude to Doron Swade for having read the proofs of this book and making some very useful suggestions.

Alan Woodward, a director of the information technology consultancy Charteris plc and a member of the British Computer Society, also read an earlier draft of the book and provided useful guidance.

David Craig, lecturer in history at Durham University, was extremely helpful in helping me to place the invention of the Jacquard loom in perspective with the progress of the British and French Industrial Revolutions.

My gratitude also to the Lyons historian who has devoted so much time and effort to the task of researching the life of Joseph-Marie Jacquard from original sources and disinterring the truth from legend, half-truth, and falsehood. He asked me not to mention him by name and I have respected his wishes.

283

I am grateful to Georges Mattelon, one of only two dozen remaining silk-weavers still using Jacquard hand-looms in Lyons, for showing me round his workshop in the city and giving me the opportunity to use an original Jacquard loom.

In addition, my thanks are due to Rolf Ziegler, who was such a fascinating guide at the House of IBM in Sindelfingen, near Stuttgart, where he introduced me to the most comprehensive collection of Hollerith punched-card machines in the world. Karl-Otto Reimers arranged for me to meet Rolf and to see the collection at Sindelfingen. I was privileged to be shown Hollerith machines in operation at the Heinz Nixdorf Museum in Paderborn, Germany, and I am grateful to Michael Mikolajczak for making this possible.

Thanks also to the staff of the Science Museum Library, London, for access to Charles Babbage's 'Scribbling Books' and to his social diary for 1844; to the staff of the Modern Manuscripts Reading Room at the British Library, London; and to Audrey Giraud and Pascale Le Cacheux, of the Museums of the History of Weaving and Decorative Arts in Lyons, for their unfailing patience and kind assistance.

The most important original source for the lives of Lord Byron, Lady Byron, and Ada Lovelace is the Lovelace-Byron Collection in the Bodleian Library, Oxford. My thanks to Colin Harris, Rob Wilkes, and Nicola Cennan for their assistance with locating material. I also thank the Earl of Lytton for granting me access to this Collection and for permission to reproduce material from it and Lady Lytton for her helpful advice. Thanks are also due to the Earl of Lytton's literary executors, Laurence Pollinger Limited.

The recently discovered letters from Ada Lovelace to Charles Babbage are held at the Northumberland Record Office. I am grateful to Sarah Cooley of the Record Office for help in this respect and to the Society of Antiquaries of Newcastle upon Tyne for its permission to view and quote from these precious letters.

I would also like to thank the following people: David Allison of the National Museum of American History; Anna Alessi; Tracy Banks; Gaëlle Baton and Marie-Chantal Mousset of the Lyons Office of Tourism; Susan Bennett of the Royal Society; Jonathan Betts of the National Maritime Museum; Marie Bouzard; Will Causton; Ben Churchill; Arthur C. Clarke; Robert Cleaver; Heni Clouts; the late I. B. Cohen; Bruce Collier; Garf Collins; Louise Coxon; Sally Day; Alison Dixon of Seaham Town Council; Beate Duncan; Edwina Ehrman of the Museum of London; Mary Essinger; Janet Foster of the Royal Statistical Society; Alan Fuller; Stephen Gillatt; Florence Greffe of the Académie des Sciences at the Institute of France; Stephen Hallett; Mike Hatton; Peter Hingley of the Royal Astronomical Society; Anthony Hyman; Louis Irwin and Bob Mann of Totnes Museum for their hospitality in Totnes and comprehensive assistance; Eddie Jephcott for assistance with translations; Friedrich W. Kistermann; Rebecca Lindskog; Jean-Yves Ligot of the Maison des Canuts in Lyons; John McCrae of Simon Langton Grammar School, Canterbury; John Rasmussen; Rebecca Salisbury; Richard K. Scher; Ginny Sennett; Diana Turner of Trinity College Library, Cambridge; Chris Weeks of the British Society of the History of Mathematics; Adrian Wilson of *Textile Month*; Michael Wright, curator of Mechanical Engineering at the London Science Museum; and Helen Yeowart of the Institute of Textiles.

My sincere thanks also to Caroline Davidson for her enthusiasm and insight, to Clive Priddle for his good sense and encouragement, to Fiona Gold for her sound advice, to my brother Rupert Essinger for his ideas and comments, and to Sheila Ableman, a queen among literary agents. Computer scientist Andrew Yeomans read this book in draft form and made many useful suggestions, as did Dr Mark George and Ian Syme, managing director of modern Jacquard loom manufacturer Stäubli UK.

Michael Rodgers of Oxford University Press was an enthusiastic advocate for the book during the OUP review process. Working with him was a delight. My great thanks also to

OUP's Marsha Filion, Abbie Headon, Jennifer Hicks, Deborah Protheroe, and Emma Simmons.

For personal reasons my thanks to Kieran Minshull of L. K. Leon & Co. and to my friends Sandy Baker and Alex Dembitz. My warmest gratitude also to Cedric Dickens, the great-grandson of Charles Dickens, for his delightful anecdotes about his early days working for the British Tabulating Machine Company.

Above all, I thank Helen Wylie for her advice, her arduous and earnest research, her careful attention to the complex task of researching the illustrations, her work on the index, and for her belief in this project from the very first day we thought of it.

James Essinger 2004

Notes on sources

In these notes, I provide details of the original sources I found most useful.

pages 7–18 Here I have drawn on original sources in Lyons, various encyclopaedia articles about sericulture, and from *A History of Textiles* by Kax Wilson (Westview Press, Boulder, Colorado, 1979) and *The Fontana History of Technology* by Donald Cardwell (Fontana Press, London, 1994).

pages 22–43 My main source for the early history of Jacquard's life are papers published in the *Bulletin Municipal Officiel* of Lyons between 1998 and 1999. As far as I am aware, these are the only reliable historical documents about Jacquard's life ever published. Other published sources, which are too often mere reworkings of existing unattested accounts of Jacquard's life, are rarely reliable.

page 46 The draft letter from Charles Babbage to Jean Arago is in the British Library, Additional Manuscripts No. 37 191, folios 287–9. The portrait of Jacquard which Babbage was finally able to obtain is in the reserve collection of the Science Museum, in London.

page 93 Jean Arago's letter explaining his problems with obtaining the Jacquard portrait for Babbage is in the British Library, Additional Manuscripts No. 37 191, folio 316.

page 58 The letter from Herschel to Babbage urging the abandonment of formality in correspondence is in Volume 2 of the Herschel Papers in the library of the Royal Society, London, folio 8.

page 78 Henry Fitton's letter of condolence to Babbage on Georgiana's death is in the British Library, London, in Additional Manuscripts 37 184, folio 80.

page 101 The evidence for when Babbage returned to Britain from Turin is inherent in a letter in Additional Manuscripts 39,191, folio 450. This is dated 11 September 1840. It was addressed to Babbage in London but redirected to an address in Ostend, where he seems to have been staying prior to coming back to Britain.

page 102 Sir Robert Peel's letter to the Earl of Haddington about the correct attitude to adopt to the financial requests of men of science is in the British Library's Additional Manuscripts 40 456, folio 98.

page 103 Peel's letter to William Buckland, showing how the Prime Minister felt about Babbage, is in Additional Manuscripts 40 514, folio 223.

pages 105–6 Henry Goulburn's letter to Babbage notifying him of the Government's decision to stop funding the Difference Engine, is in Additional Manuscripts V 37 192, folios 172–3.

pages 107–12 Babbage's detailed account of his abortive meeting with Sir Robert Peel on Friday, 11 November 1842 is in Additional Manuscripts 37 192, folios 189–93.

page 112 Although I do not mention this directly in the text, for details of a dinner-party given by Dickens which Babbage and Lord and Lady Lovelace attended see *The Letters of Charles Dickens*, Vol. *5, 1847–1849*, p. 513, ed. Storey/Fielding (OUP, Oxford, 1981).

page 112 For Dickens's letter to Henry Austin about the bill for Tavistock Place, see *The Letters of Charles Dickens*, Vol. 6, p. 556, ed. Storey/Tillotson and Burgis (OUP, Oxford, 1988).

pages 128–30 On the subject of Babbage's personal life, there is an intriguing letter to him from a Reverend Lunn in the British Library's Additional Manuscripts 37 185, folio 310. This suggests that Babbage had asked Lunn to enquire about a certain lady to see whether she might be a suitable candidate for a wife for Babbage. This seems to constitute solid evidence that Babbage had not necessarily resigned himself to permanent bachelor life after Georgiana's death.

Bibliography

Ackroyd, Peter. *Dickens*. London: Sinclair-Stevenson Limited, 1990.

Aiken, Howard. *A Manual of Operation for the Automatic Sequence Controlled Calculator*. Cambridge, Massachusetts: Harvard University Press, 1946.

Arizzoli-Clémentel, Pierre. *The Textile Museum, Lyons*. Paris: Musées et Monuments de France/Fondation Paribas, 1996.

Aspray, William (Editor). *Computing Before Computers*. Ames, Iowa: Iowa State University Press, 1990.

Austrian, Geoffrey D. *Herman Hollerith, Forgotten Giant of Information Processing*. New York: Columbia University Press, 1982.

Babbage, Charles. *On the Economy of Machinery and Manufactures*. London: Charles Knight, 1832.

Babbage, Charles. *On the Principles and Development of the Calculator*. New York: Dover Publications, 1961.

Babbage, Charles. *Passages from the Life of a Philosopher*. New Brunswick, New Jersey: Rutgers University Press and Piscataway, New Jersey: IEEE Press, 1994.

Babbage, Charles. *Science and Reform, Selected Works of Charles Babbage*. Cambridge: Cambridge University Press, 1989.

Barret-Ducrocq, Françoise. *Love in the Time of Victoria*. London: Penguin Books, 1992.

Benoit, Bruno and Curtet, Raymond. *Quand Lyon rugit ... Les Colères de Lyon du Moyen Âge à nos Jours.* Lyon: Editions Lyonnaises d'Art et d'Histoire, 1998.

Berners-Lee, Tim. *Weaving the Web.* London: Orion Business, 1999.

Black, Edwin. *IBM and the Holocaust.* New York and London: Crown Publishers/Little Brown & Co., 2001.

Bouzard, Marie. *La Soierie Lyonnaise du XVIIIe au Xxe siècle.* Lyon: Editions Lyonnaises d'Art et d'Histoire, 1997.

Briggs, Asa. *The Age of Improvement 1783–1867.* Harlow, Essex: Addison Wesley Longman, 1979.

Bromley, Allan. *The Babbage Papers in the Science Museum.* London: The Science Museum, 1991.

Brown, Donald. *Charles Babbage—The Man and his Machine.* Totnes: The Totnes Museum Study Centre, 1992.

Brown, Janet. *Charles Darwin—Voyaging.* London: Pimlico, 1996.

Buxton, H.W. *Memoir of the Life and Labours of the Late Charles Babbage Esq. F.R.S..* Cambridge, Massachusetts: MIT Press and Tomash Publishers (Los Angeles and San Francisco), 1988.

Campbell-Kelly, Martin and Aspray, William. *Computer: A History of the Information Machine.* New York: HarperCollins, 1986.

Cardwell, Donald. *The Fontana History of Technology.* London: Fontana Press, 1994.

Ceruzzi, Paul E. *A History of Modern Computing.* Cambridge, Massachusetts: MIT Press, 1998.

Clarke, Arthur C. *Profiles of the Future.* London: Gollancz, 1982.

Cohen, I. Bernard. *Howard Aiken: Portrait of a Computer Pioneer.* Cambridge, Massachusetts: MIT Press, 1998.

Collier, Bruce. *The Little Engines that Could've.* New York and London: Garland Publishing, 1990.

Cronin, Vincent. *Napoleon.* London: HarperCollins, 1994.

Desmond, Adrian and Moore, James. *Darwin.* London: Penguin Books, 1992.

Dickens, Charles. *A Tale of Two Cities.* London: Chapman & Hall, 1860.

Dickens, Charles. *Little Dorrit.* London: Chapman & Hall, 1855.

Eames, Charles and Ray. *A Computer Perspective—Background to the Computer Age.* Cambridge, Massachusetts: Harvard University Press, 1990.

Enriquez, Juan. *As the Future Catches You.* New York: Random House, 2001.

Étèvenaux, Jean. *Charles-Marie Jacquard.* Lyon: LUGD, 1994.

Eymard, Paul. *Historique de Métier Jacquard.* Lyons: Imperial Society of Agriculture, Natural History and the Useful Arts, 1863.

Gibson, William and Sterling, Bruce. *The Difference Engine.* London: Vista, 1996.

Grosskurth, Phyllis. *Byron, The Flawed Angel.* London: Hodder and Stoughton, 1997.

Hibbert, Christopher. *The French Revolution.* London: Penguin Books, 1980.

Hobsbawm, Eric. *The Age of Revolution 1789–1848.* London: Abacus, 1999.

Howse, Derek. *Greenwich Time and the Longitude*. London: Philip Wilson Publishers, 1997.

Hyman, Anthony. *Charles Babbage, Pioneer of the Computer*. Oxford: Oxford University Press, 1982.

Jacquemin, L. *Discover Lyon*. Bourg-en-Bresse, France: Editions de la Taillanderie, 1998.

Jacquemin, L. *Visiting the Traboules of Lyon*. Bourg-en-Bresse, France: Editions de la Taillanderie, 1992.

Johnson, Nicola. *Eighteenth Century London*. London: HMSO, 1991.

Kaplan, Fred. *Dickens—A Biography*. London: Hodder and Stoughton, 1988.

King-Hele, D. G. (Editor). *John Herschel 1792–1871: A Bicentennial Commemoration*. London: The Royal Society, 1992.

Lardner, Dionysius. *The Cabinet Cyclopædia—Treatise on the Origin and Present State of the Silk Manufacture*. London: Longman, Rees, Orme, Brown, and Green, 1831.

Lewis, Gwynne. *The French Revolution—Rethinking the Debate*. London: Routledge, 1993.

McLynn, Frank. *Napoleon—A Biography*. London: Pimlico, 1998.

Metropolis N., Howlett J., Gian-Carlo Rota (Editors). *A History of Computing in the Twentieth Century*. London and New York: Academic Press, 1980.

Moore, Doris Langley. *Ada: Countess of Lovelace: Byron's Legitimate Daughter*. London: John Murray, 1977.

Moseley, Maboth. *Irascible Genius: A Life of Charles Babbage, Inventor*. London: Hutchinson, 1964.

Parker, Rozsika. *The Subversive Stitch*. London: The Women's Press, 1984.

Roberts, J. M. *The French Revolution*. Oxford: Oxford University Press, 1997.

Shurkin, Joel. *Engines of the Mind*. New York and London: W. W. Norton, 1994.

Spufford, Francis and Uglow, Jenny. *Cultural Babbage: Technology, Time and Invention*. London: Faber and Faber, 1996.

Standage, Tom. *The Victorian Internet*. New York: Walker Publishing, 1998.

Stein, Dorothy. *Ada: A Life and Legacy*. Cambridge, Massachusetts: MIT Press, 1985.

Sutherland, D.M.G. *France 1789–1815: Revolution and Counterrevolution*. London: Fontana Press, 1985.

Swade, Doron. *Charles Babbage and his Calculating Engines*. London: The Science Museum, 1991.

Swade, Doron. *The Cogwheel Brain*. London: Little, Brown, and Company, 2000.

Tocqueville de, Alexis. *The Ancien Régime*. London: J. M. Dent & Sons, 1988.

Toole, Betty. *Ada, the Enchantress of Numbers*. Mill Valley, California: Strawberry Press, 1992 .

Tweedale, Geoffrey. *Calculating Machines and Computers*. London: Shire Publications, 1990.

Wilson, Kax. *A History of Textiles*. Boulder, Colorado: Westview Press, 1979.

Woolley, Benjamin. *The Bride of Science*. London: Macmillan, 1999.

Index

Sub-entries are in chronological order. *Italic* numbers denote references to illustrations. HA = Howard Aiken, CB = Charles Babbage, HH = Herman Hollerith, J = Joseph-Marie Jacquard, AL = Ada Lovelace.